IMAGES
of America

LOUDOUN COUNTY FIRE AND RESCUE APPARATUS HISTORY

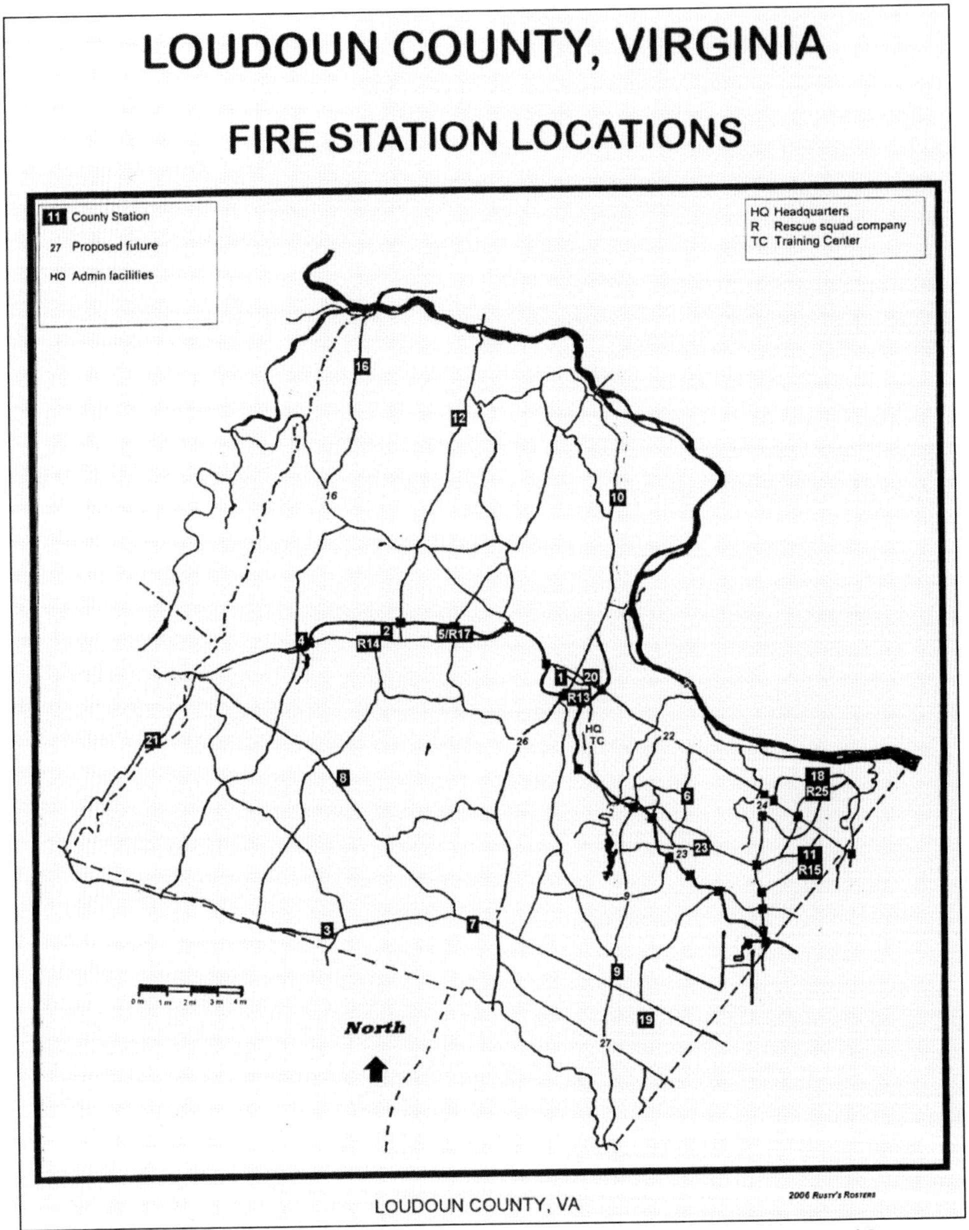

Pictured is a map of the fire and rescue departments of Loudoun County. (Courtesy of Rusty Gillman.)

On the Cover: The Ashburn Volunteer Fire Department (VFD) was first organized in 1944 after a house fire near the Ashburn community claimed the lives of several children. As a result of this tragedy, the community came together to form a fire department. The fire station was built on Ashburn Road, and over the years, Station No. 6 has been modified to include additional bays and a social hall. Members of the Ashburn Volunteer Fire Department were photographed in front of their 1948 GMC fire truck, which was built by the American Fire Apparatus Company of Battle Creek, Michigan. Behind the garage doors sits Ashburn's first fire truck, a 1928 American LaFrance. (Courtesy of Ashburn Volunteer Fire and Rescue Department.)

IMAGES
of America

LOUDOUN COUNTY FIRE AND RESCUE APPARATUS HISTORY

Mike Sanders

ARCADIA
PUBLISHING

ISBN 978-0-7385-5263-7

Published by Arcadia Publishing
Charleston SC, Chicago IL, Portsmouth NH, San Francisco CA

Printed in the United States of America

Library of Congress Catalog Card Number: 2007922548

For all general information contact Arcadia Publishing at:
Telephone 843-853-2070
Fax 843-853-0044
E-mail sales@arcadiapublishing.com
For customer service and orders:
Toll-Free 1-888-313-2665

Visit us on the Internet at www.arcadiapublishing.com

Dedicated to all of the men and women in Loudoun County, Virginia, who have given their time and energy in providing fire and rescue services to their communities.

Contents

ACKNOWLEDGMENTS

I have been told that little boys love fire trucks. That was the case for me. As a boy, I used to pedal my bike at daring speeds to see a fire truck responding. When I owned my first Instamatic camera, my photographs were of fire trucks. Over the years, I have visited many Virginia fire stations, and it has expanded my hobby in fascinating ways. Now I have the awesome responsibility of driving fire trucks in my community—and the love has never gone away.

This book was a labor of love, and I have many people to thank. Specifically I want to thank each of the departments in Loudoun County that offered their old photographs and provided unique information so that this history could be captured. Many people provided additional help with photographs. There are too many names to list here, but their contributions are acknowledged with the photographs they provided. I do want to specifically acknowledge three people: Rusty Gillman, Pete West, and Robert Kimball. Rusty is a walking history book on Washington, D.C., metropolitan area fire departments. His Loudoun County information was extremely valuable during this research. Thank you, Rusty. Pete worked with me on a previous book project, and I learned a great deal from him during that time. Pete also shared with me many Loudoun County photographs from his collection. Thank you, Pete. Robert shared with me old photographs he took of Loudoun County, which were very useful. Thank you, Robert, for your help. Finally, unless otherwise noted, all images are from the author's collection.

Taking on this project meant that I had to juggle my schedule with my family. The Lord has blessed me with my greatest joy in my life, Lisa and my five sons. Andrew, David, Jordan, Nick, and Luke—thank you guys for allowing me to spend time working on this project. Although I love the time I spend with my hobby, I love you guys so much more. Thanks for making me so proud. Most of all, thank you, Lisa, for your support and encouragement. I love and admire you and wonder how I deserved such a gift.

INTRODUCTION

Loudoun County, Virginia, is one of the fastest growing counties within the United States. For many years, agriculture was the primary business opportunity for county residents. The majority of the county consisted of numerous farms, homes, and small businesses.

When Dulles International Airport opened in the late 1960s, many new businesses began to appear in the county, and communities were built to house the many families that began working in Loudoun County. Today numerous high-tech businesses call Loudoun County home. Much of the eastern part of the county has many new homes and office parks. The western part of the county has limited growth, although new communities have begun to appear over the years. The county can best be described as split—the eastern half is suburban. The western end of the county is more rural, and many people are fighting to keep its "quaint" look and feel. However, growth continues to move toward the western end of the county.

The Loudoun County fire and rescue departments consist of a combination of separate and independent companies that specialize in fire, rescue, or both. Today there are a total of 17 volunteer fire and rescue departments. No matter their specialty, all departments respond to fire and rescue emergencies. Working in conjunction with these volunteer fire and rescue organizations is the Loudoun County Department of Fire, Rescue, and Emergency Management. Career fire and rescue personnel are hired and managed by this county organization.

The history of fire and rescue apparatus in Loudoun County is one that could be found in many fire and rescue departments across our country. Fire and rescue departments were established because the community recognized the need for this service; however, gathering all of the necessary equipment to begin a fire or rescue department was challenging. The residents of Loudoun County supported these initiatives and donated the funds to help get started. In addition, numerous businesses donated vehicles to be used as ambulances or rescue trucks. Much of the early apparatus in Loudoun County were trucks that were converted for firefighting or medical care. Some departments purchased used fire apparatus from other departments to help establish fire and rescue protection within their communities. As departments became more established and growth continued, new apparatus have been purchased to provide state-of-the-art fire protection in Loudoun County. You will also see how some of Loudoun County's fire and rescue stations have preserved the past by continuing to keep older trucks that once served their communities but are no longer used for firefighting or medical care. There are a variety of images of older apparatus still currently at home in Loudoun County. This collection of photographs contains a variety of old, new, and unusual fire and rescue apparatus that have served the residents of Loudoun County.

One challenge I faced when completing this book was how to divide the chapters. My goal was to give the reader a glimpse of the fire and rescue apparatus history in the order in which the station numbers were established. In addition, by station, I wanted to show the reader the history of the fire and rescue trucks from some of the earliest purchases to some of today's fire trucks. I believe I have accomplished that goal.

Three major chapters were developed for this publication. Chapter one, "The Early Years," covers the county's first fire and rescue stations. Included in this chapter are the apparatus histories of the following:

Leesburg Volunteer Fire: Station Nos. 1 and 20
Purcellville Volunteer Fire: Station No. 2
Middleburg Volunteer Fire and Rescue: Station No. 3
Round Hill Volunteer Fire and Rescue: Station No. 4
Hamilton Volunteer Fire: Station No. 5
Ashburn Volunteer Fire and Rescue: Station Nos. 6 and 23

Chapter two, "Loudoun County Growth," captures the fire and rescue stations that were built and organized as Loudoun County began to experience a time of growth. The fire and rescue departments covered in chapter two include the following:

Aldie Volunteer Fire: Station No. 7
Philomont Volunteer Fire: Station 8
Arcola Pleasant Valley Volunteer Fire and Rescue: Station Nos. 9 and 19
Lucketts Volunteer Fire: Station No. 10
Sterling Volunteer Fire: Station Nos. 11 and 18
Lovettsville Volunteer Fire and Rescue: Station No. 12
Loudoun County Volunteer Rescue Squad: Station No. 13
Purcellville Volunteer Rescue Squad: Station No. 14
Sterling Volunteer Rescue Squad: Station Nos. 15 and 25
Neersville Volunteer Fire and Rescue: Station No. 16
Hamilton Volunteer Rescue Squad: Station No. 17

Finally chapter three, "Loudoun County Department of Fire, Rescue, and Emergency Management," captures the apparatus purchased by the Loudoun County government, which are used in conjunction with those from all volunteer fire and rescue organizations in the county.

Fire departments were assigned a station number when they became incorporated or chartered. During the 1970s, station numbers were assigned to the volunteer rescue squads. Some Loudoun County government–purchased apparatus received a "50" designation, and training-center apparatus received the "90" designation. Other station numbers were assigned to fire departments that were not a part of the Loudoun County system but ran mutual aide calls with Loudoun. Station No. 21 is Mount Weather, and Station No. 22 is the Xerox fire brigade.

My greatest challenge during this project was deciding which photographs to include and which images I would have to leave out of the book. That process became quite a challenge, and I have made numerous revisions to the photograph layout when making those decisions. My goal was to provide many interesting apparatus images that would make the reader enjoy this book, and I feel that I have accomplished that goal.

Although I have many Loudoun County fire and rescue apparatus photographs within my collection, one of the most exciting elements of this project has been the discovery of old apparatus photographs located at the stations or in an individual's private collection. The willingness of numerous people to share this historical information will be valued for many years to come.

The information contained in this book was developed from numerous sources and is as accurate as possible. I would greatly appreciate hearing from readers with any revisions or additional information. I hope the reader will enjoy his journey in viewing a representation of the fire and rescue apparatus that have protected and continue to protect Loudoun County residents.

One

The Early Years
Station Nos. 1–6

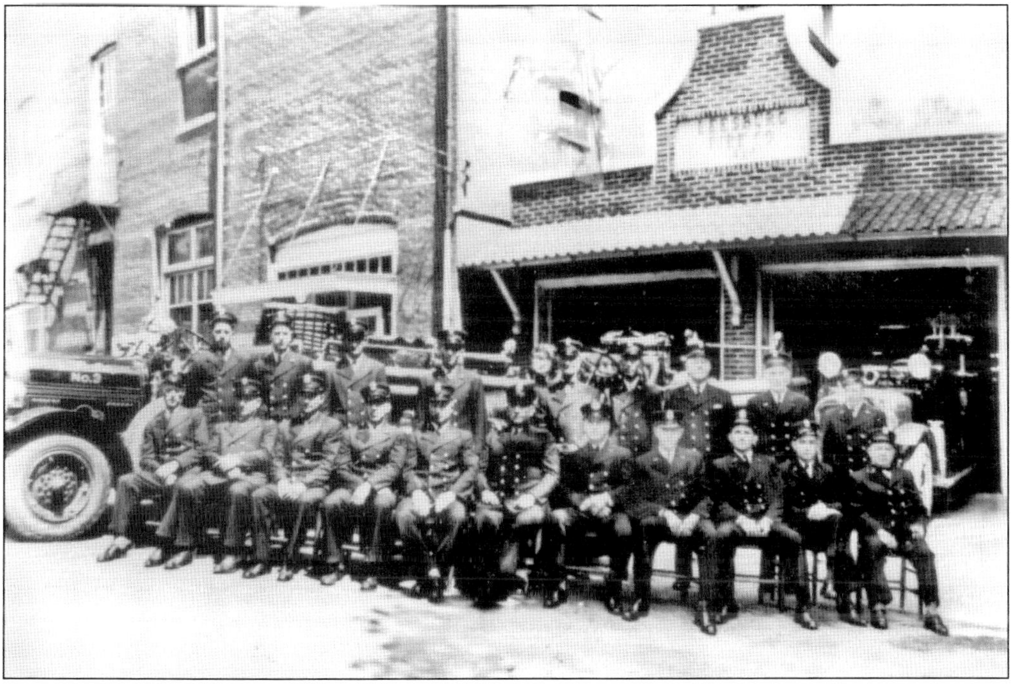

In the late 1820s, Leesburg's first bucket brigade was formally organized as the Leesburg Star Fire Company. In 1863, the Star Fire Company was reorganized and was named the Leesburg Volunteer Fire Company. In August 1927, the department moved into a new station on Loudoun Street, as shown here. These 20 firefighters were photographed in front of the department's 1929 Seagrave (left) and a 1925 Seagrave (right). Behind them is a 1916 Dodge truck that had its frame extended to carry ladders. The sign on the station reads, "Leesburg Fire Co. 1927." During this time period, all calls six miles or more out of the town limits received two blasts of the house siren, and the department had to receive permission from either the mayor or town officials to respond outside the Leesburg limits. One bylaw from 1925 said that the "driver was to slow down to 15 mph when seeing a member to allow him to mount the engine. Under no condition was he allowed to come to a complete stop." (Courtesy of J. B. Anderson.)

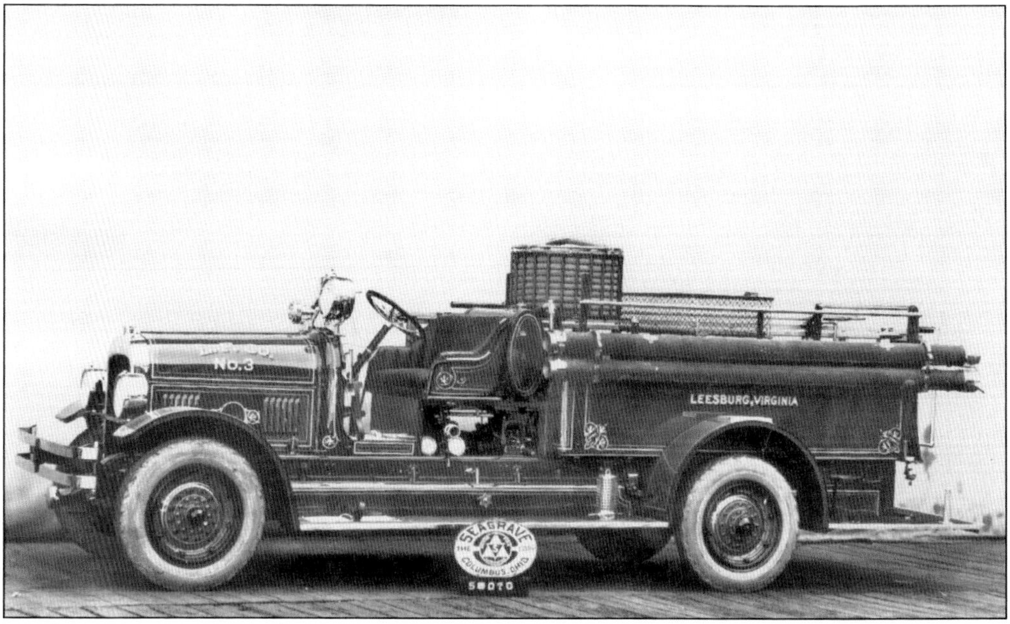

Leesburg's 1929 Seagrave has a 500-gallon-per-minute pump and carries 125 gallons of water. The fire department membership went to the Leesburg Town Council requesting support for a new fire truck. The town council agreed and paid for half of the truck. The photograph above is the Seagrave delivery photograph, taken just after Seagrave in Wisconsin built the truck. The Seagrave arrived by train at Leesburg's old train station in August 1929 and is still owned today by the fire department after an extensive renovation. Today the Seagrave is taken to a variety of public events. The lower photograph was taken at the 2006 Apple Blossom Festival in Winchester, Virginia, during the firefighters' parade. (Above, courtesy of J. B. Anderson.)

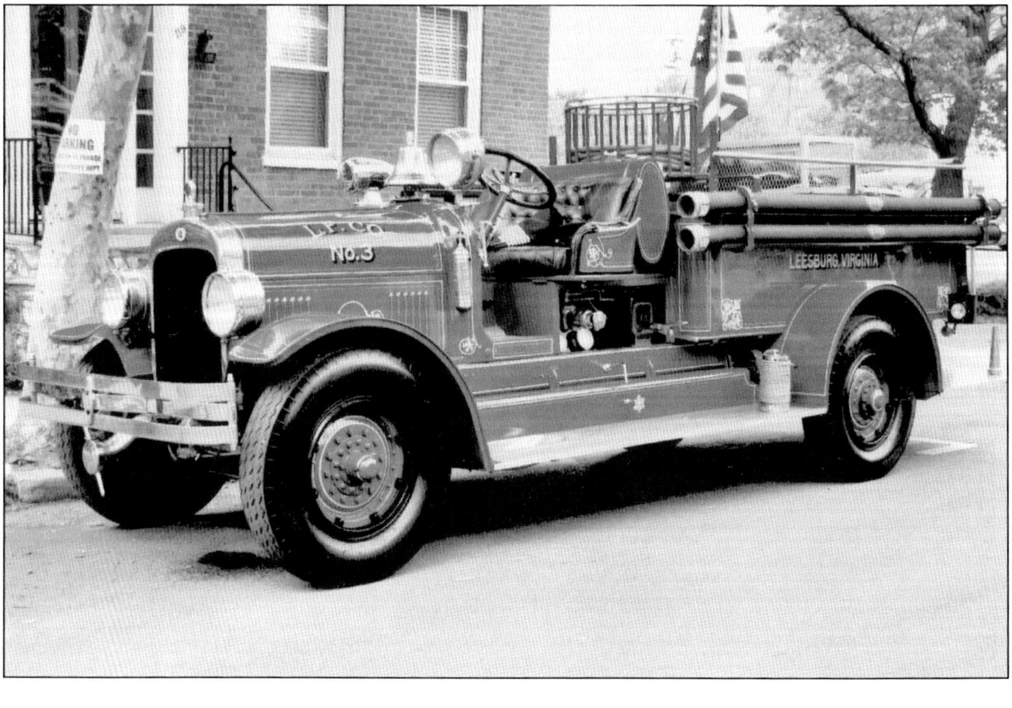

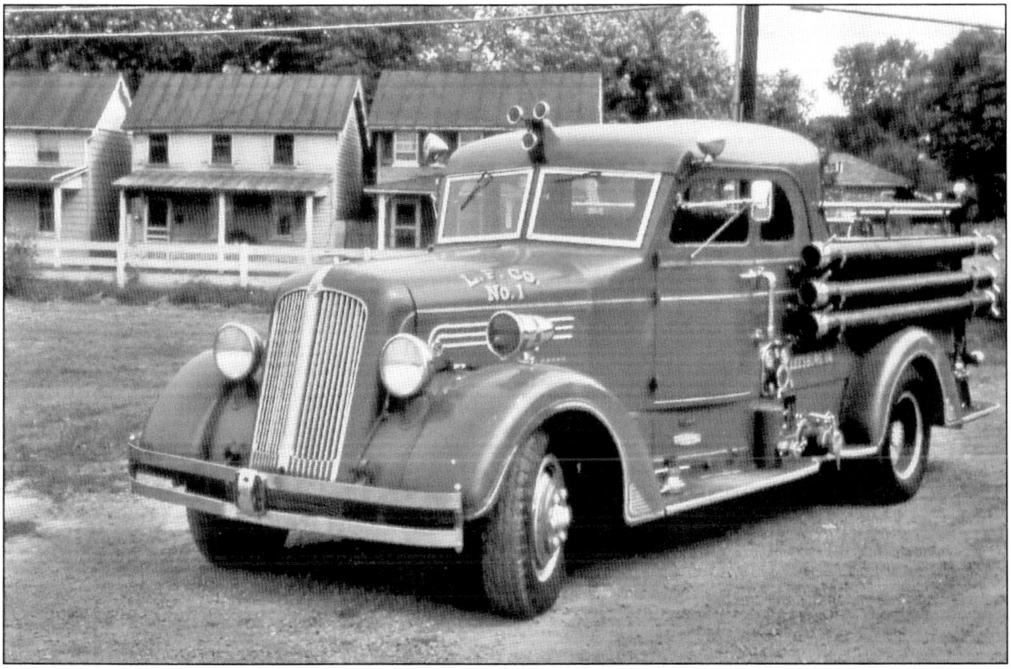

Leesburg purchased this 500-gallon-per-minute Seagrave engine in 1937. Note the very small pump panel that is located to the right of the driver's door. A small Roto-Ray warning light is mounted on the top of the cab roof. Today the 1937 Seagrave is privately owned in Leesburg. (Courtesy of David Moltrup and Mark Baker.)

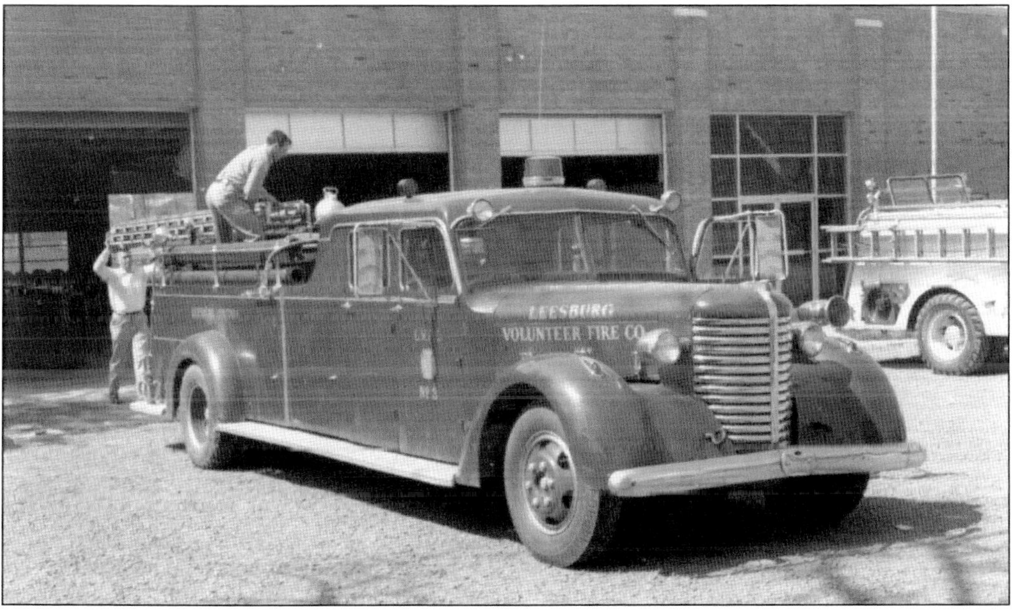

The Buffalo Fire Apparatus Company built fire trucks in Buffalo, New York. In 1946, Leesburg purchased a Buffalo complete with a 500-gallon-per-minute pump. This four-door fire truck was ahead of its time. Most fire trucks had open cabs that exposed firefighters to the rain and cold. Leesburg's Buffalo provided firefighters protection from the elements. This late-1960s photograph shows the Buffalo at the fire station at 215 Loudoun Street.

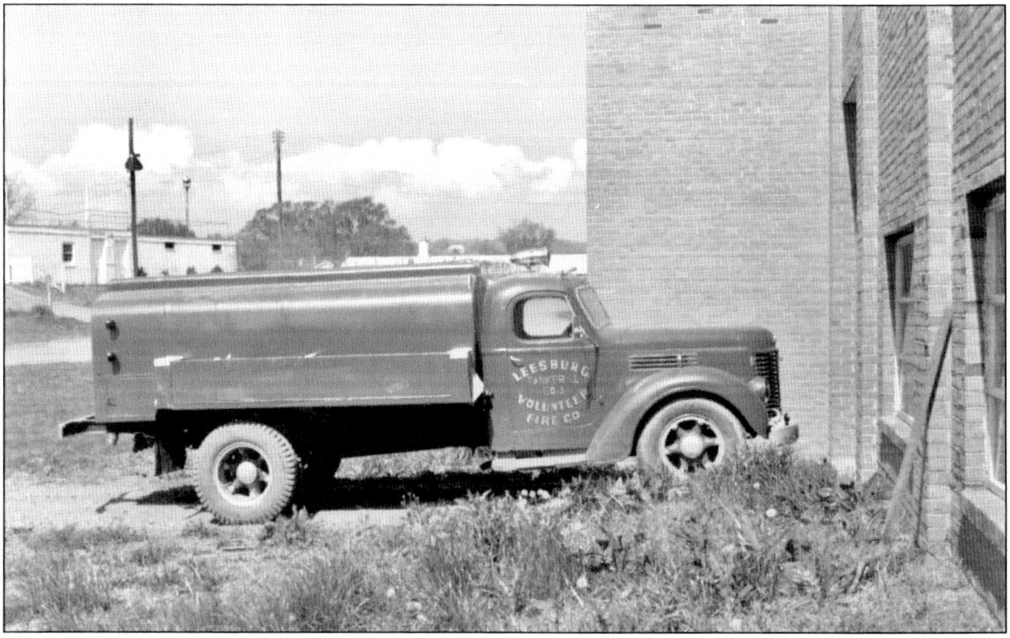

Because Leesburg has areas with no water hydrants, tanker trucks have been an important part of the department's history. One of Leesburg's earlier tankers was this old International fuel truck previously owned by O. G. Hall Fuel and built in the late 1940s. The tanker truck, photographed behind the station at 215 Loudoun Street, carried 1,200 gallons of water and was nicknamed "Old Maude" by the Leesburg volunteers. The tanker would be rebuilt in 1964.

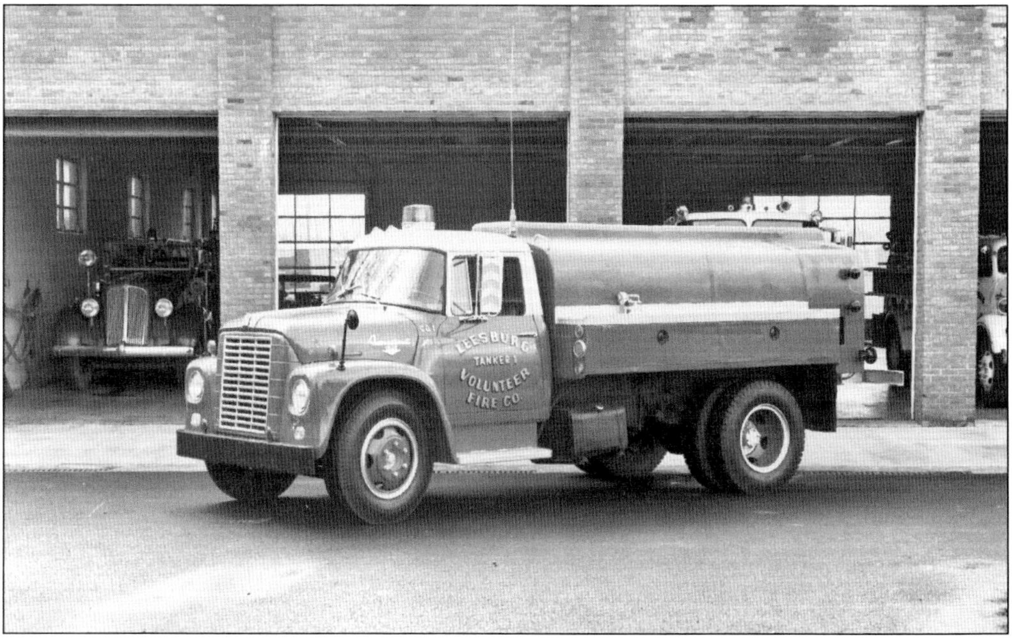

Leesburg's old International tanker needed to be repowered. In 1964, the old International cab was replaced with a new International cab. The same 1,200-gallon tanker body was utilized for Tanker No. 1. It was later sold to the Blue Ridge Volunteer Fire Department in Clarke County. The tanker was photographed in front of Station No. 1 on Loudoun Street. (Courtesy of J. B. Anderson.)

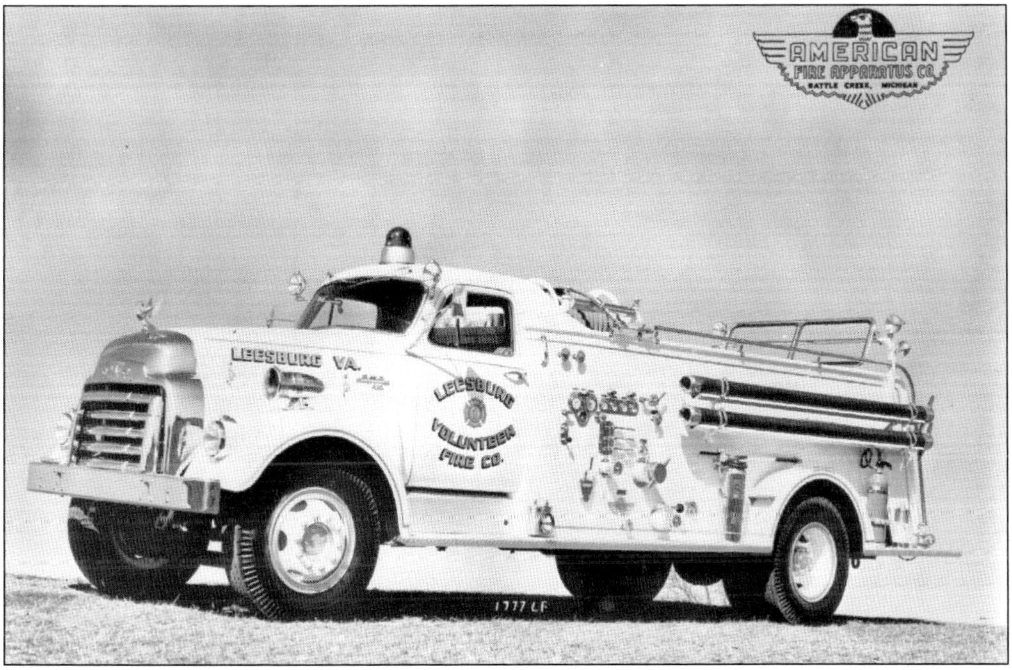

Many fire-apparatus manufacturers photograph a new fire truck just prior to its delivery. Such was the case for Leesburg's new 1953 GMC, which was built by the American Fire Apparatus Company of Battle Creek, Michigan. The fire truck was ordered through the Glenn D. Culbert Company, a fire-truck distributor located in the Washington, D.C., area and popular with many departments. While earlier Leesburg fire trucks were painted red, the new GMC arrived in its all-white color. It would, however, later be repainted to red and white. The GMC came equipped with a 500-gallon-per-minute pump, 500 gallons of water, and a rear windshield to help protect the firefighters from the elements. (Below, courtesy of Chuck Madderom.)

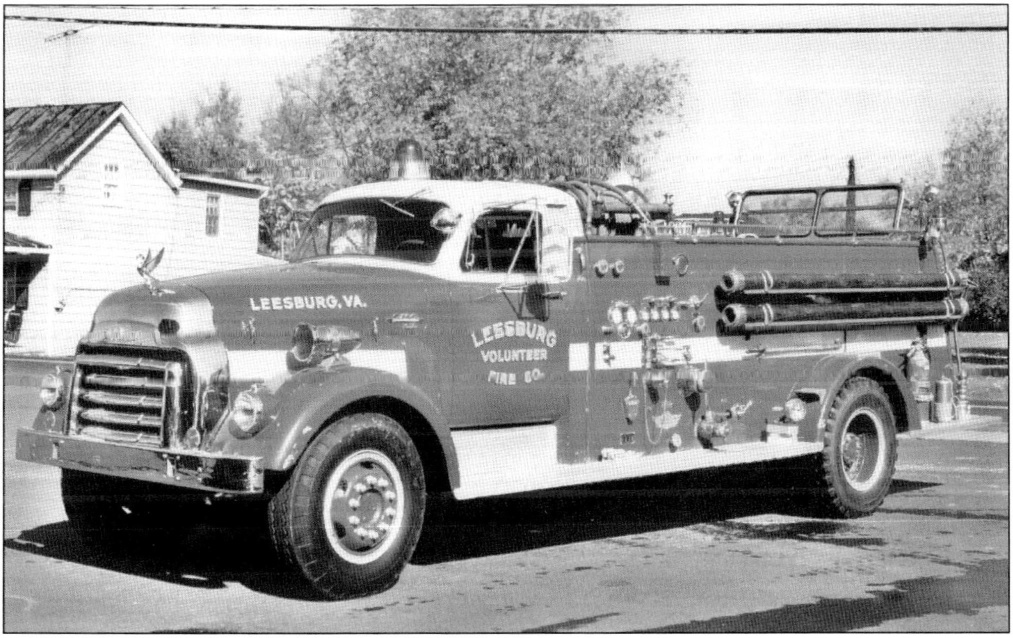

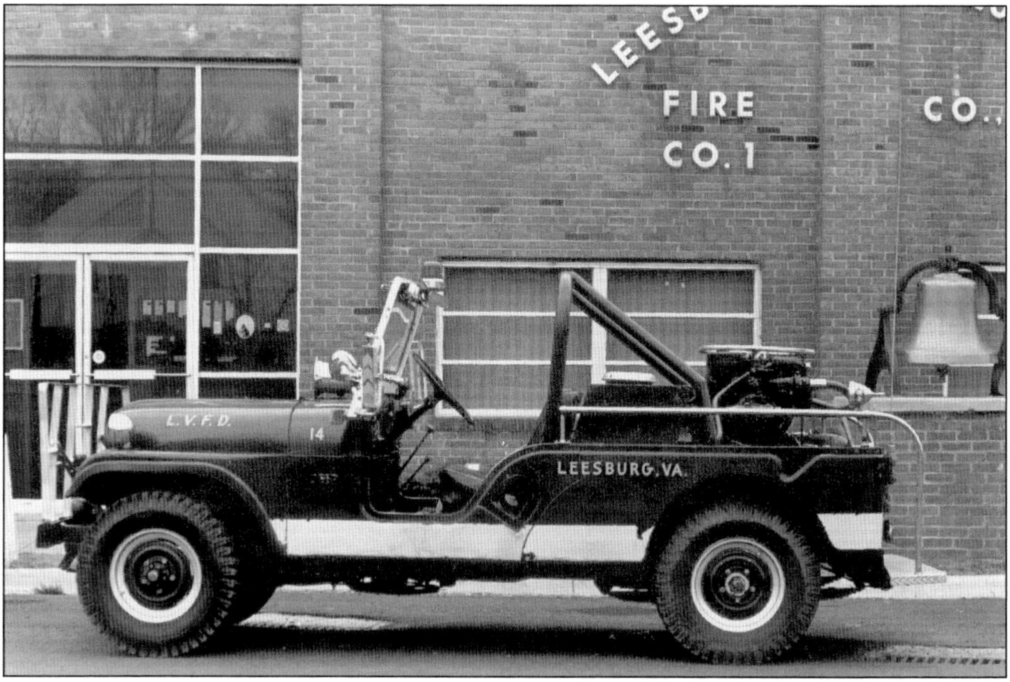

The Jeep has been a common brush-fire unit during Leesburg's history. Station No. 1 originally operated a 1953 Jeep. It was replaced in 1964 when Leesburg purchased this Jeep. Photographed in front of Station No. 1, the Jeep had a grab rail along the side from Leesburg's 1929 Seagrave engine. Leesburg operated other Jeeps as brush-fire units, including a 1977 model and a 2006 Jeep. (Courtesy of Leesburg VFD.)

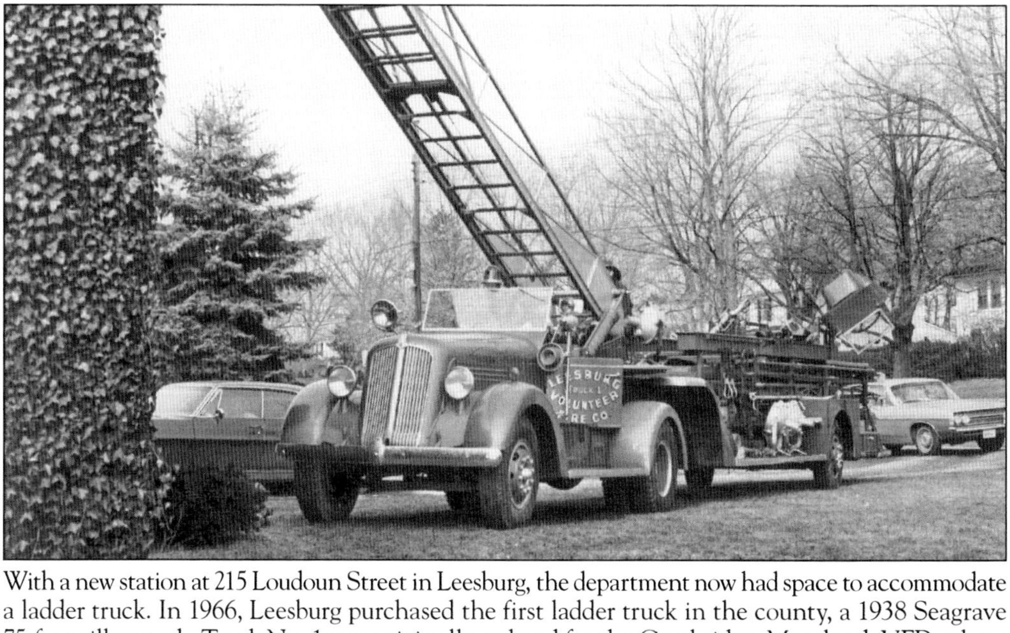

With a new station at 215 Loudoun Street in Leesburg, the department now had space to accommodate a ladder truck. In 1966, Leesburg purchased the first ladder truck in the county, a 1938 Seagrave 75-foot tiller truck. Truck No. 1 was originally ordered for the Cambridge, Maryland, VFD, where it was nicknamed "Queen Mary." To raise the ladder, the tillerman's seat would lift up and move to the side. The 1938 Seagrave tiller truck was replaced in 1973. (Courtesy of J. B. Anderson.)

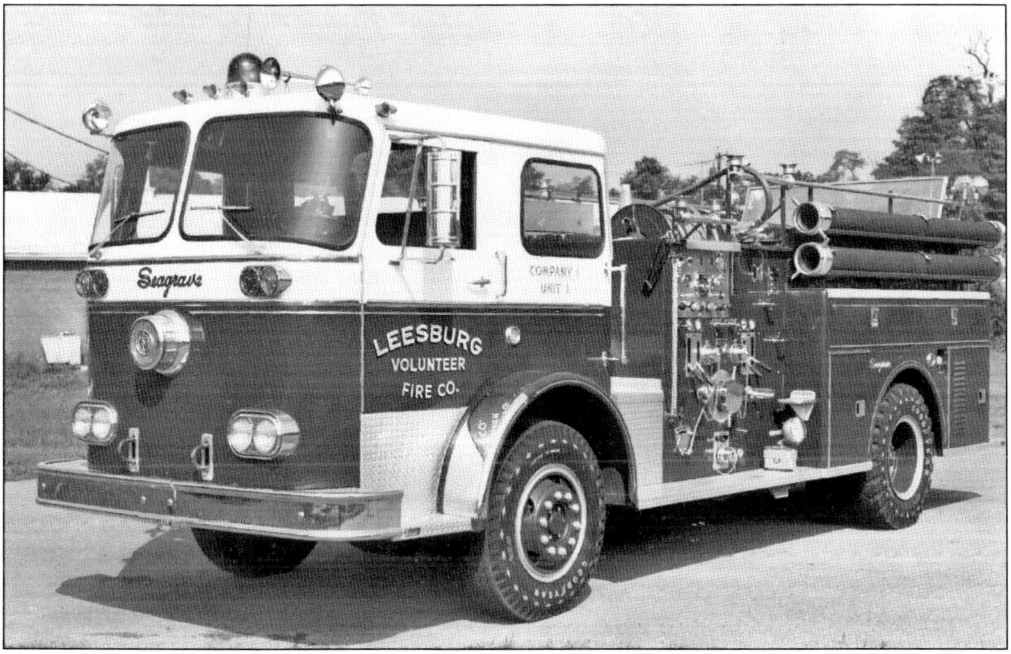

Leesburg has the distinction of owning the first diesel-powered fire truck in Loudoun County. The department took delivery of this red-and-white 1968 Seagrave, which was equipped with a 1,000-gallon-per-minute pump. It would be rehabbed during the mid-1970s by 4-Guys, Inc., a fire-truck manufacturer in Pennsylvania. The Seagrave was sold in 1990 to the Round Hill, Virginia, VFD, which is Station No. 4 in Loudoun County. (Courtesy of J. B. Anderson.)

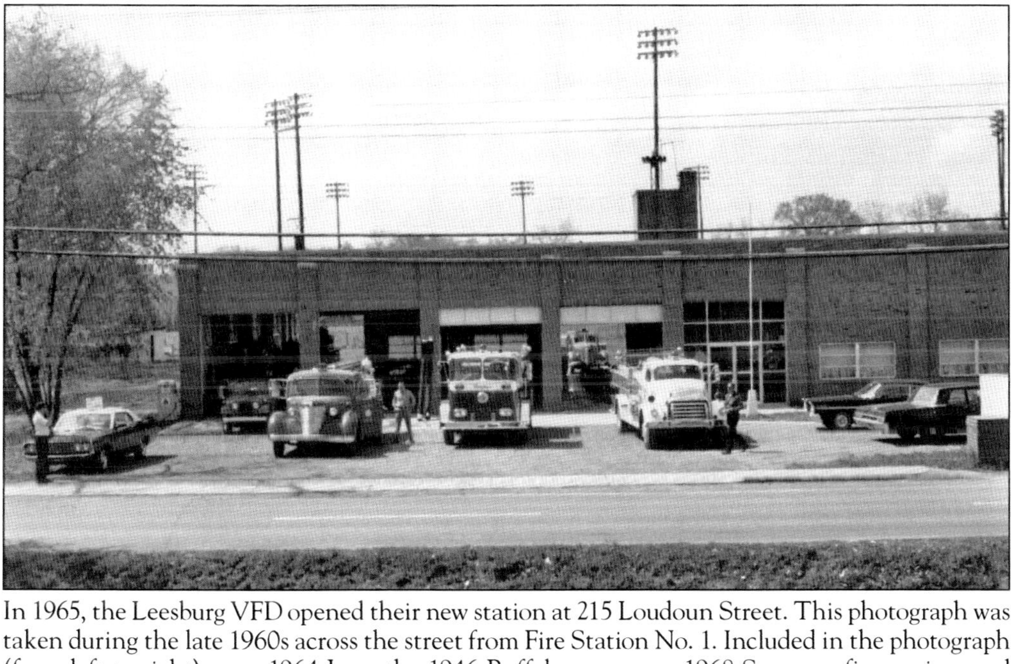

In 1965, the Leesburg VFD opened their new station at 215 Loudoun Street. This photograph was taken during the late 1960s across the street from Fire Station No. 1. Included in the photograph (from left to right) are a 1964 Jeep, the 1946 Buffalo pumper, a 1968 Seagrave fire engine, and the white 1954 GMC/American. Tanker No. 1, the old International, is sitting out back, where a ballpark and swimming pool were located.

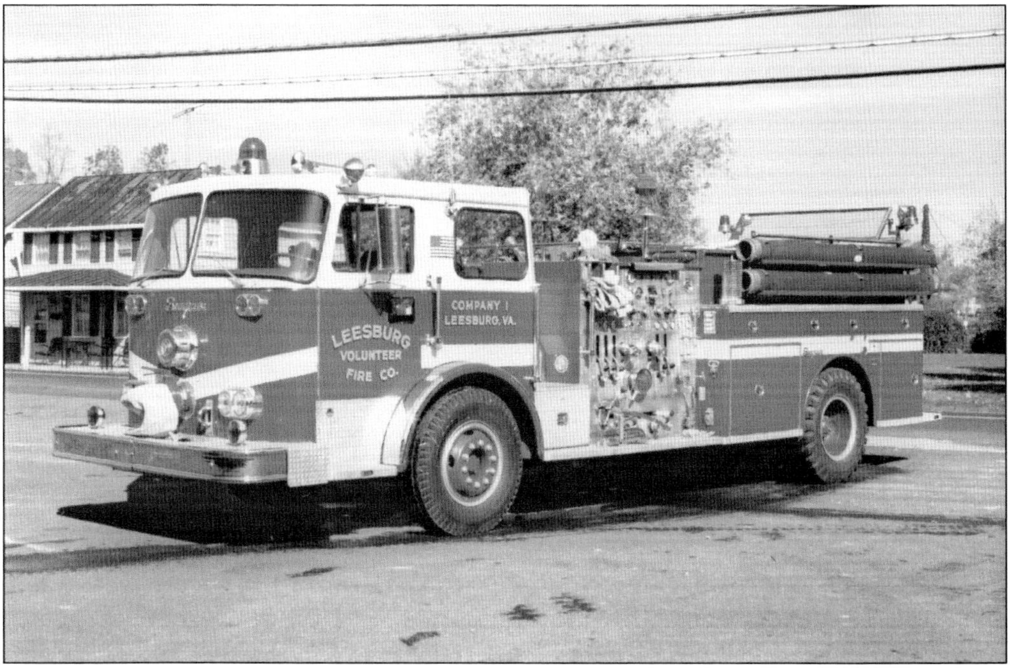

For many years, fire trucks built by Seagrave Fire Apparatus of Clintonville, Wisconsin, were the fire trucks of choice for Leesburg. In 1973, Seagrave delivered this fire engine, complete with a 1,250-gallon-per-minute pump and a 750-gallon water tank. When the 1973 Seagrave was delivered, it originally had a windshield on the back step to protect firefighters from the elements. In 1988, the department had the Seagrave returned to Clintonville and completely refurbished, including the addition of more compartment spaces to carry extra equipment. Leesburg still owns the Seagrave today, and it runs as a reserve engine. (Above, courtesy of Chuck Madderom.)

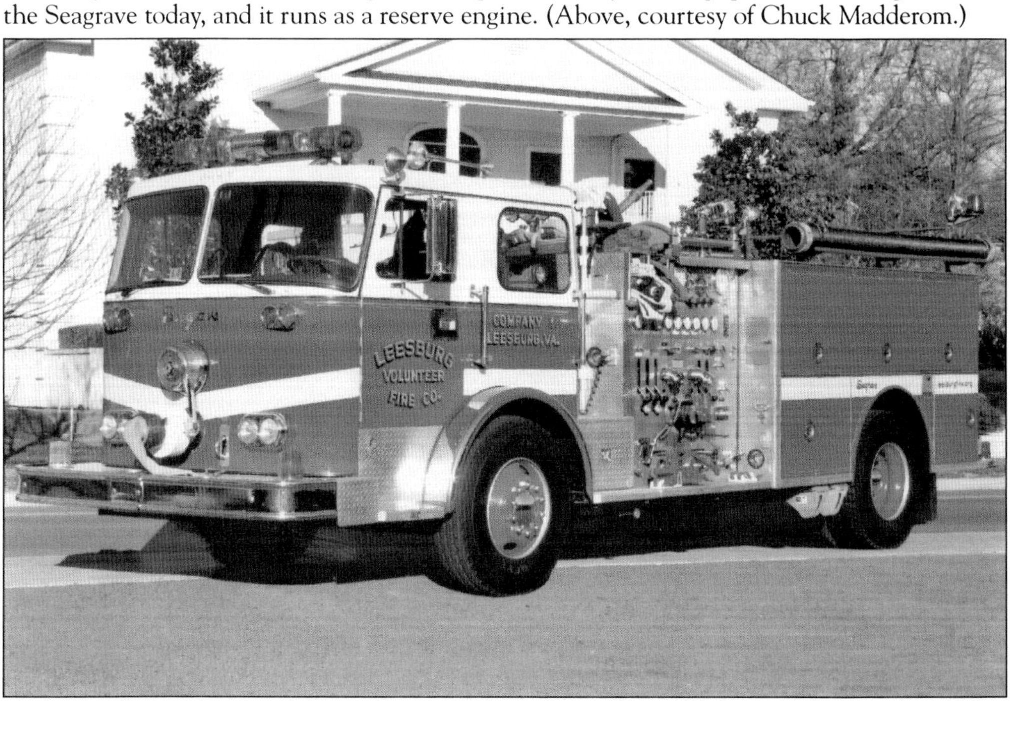

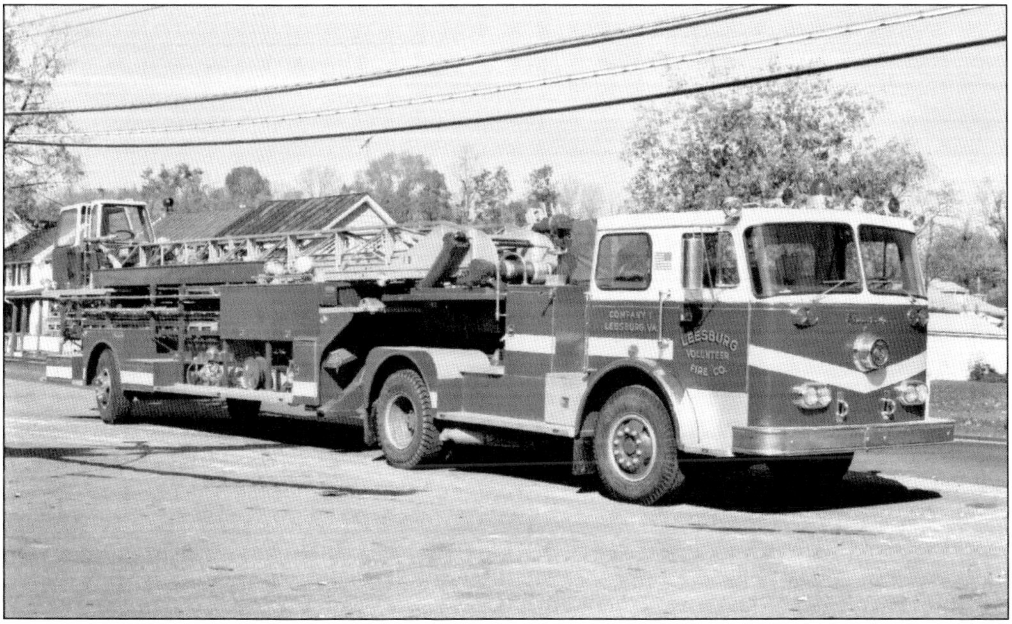

The Leesburg Volunteer Fire Department replaced their 1938 Seagrave ladder truck with a new Seagrave 100-foot tiller in 1973. The tiller truck proved valuable because of Leesburg's narrow downtown streets. After serving nearly 20 years, the Seagrave tiller (along with another ladder truck) was sold to Baltimore City, Maryland, which was in desperate need of ladder trucks. It served Baltimore as both a front-line and reserve ladder truck. (Courtesy of Chuck Madderom.)

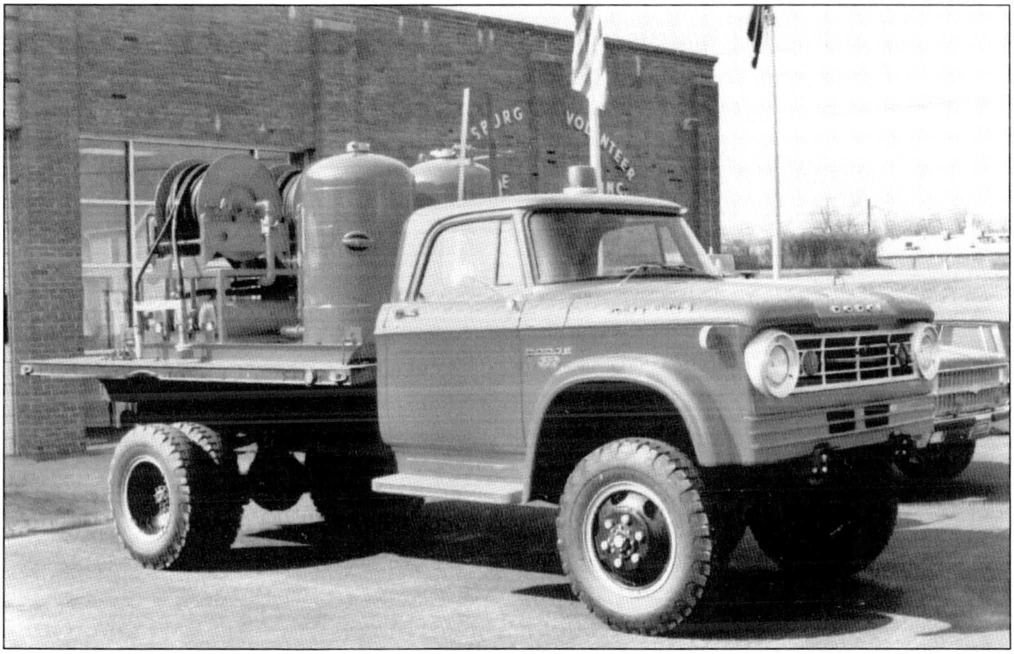

Leesburg is home to the Leesburg Executive Airport, which was built in 1963. Originally known as Godfrey Field, the land was owned by radio and television personality Arthur Godfrey, who used a DC-3 airplane. The airport donated this 1970s Dodge airport crash truck to Leesburg to be used as the primary protection for aircraft and other emergencies at the airport.

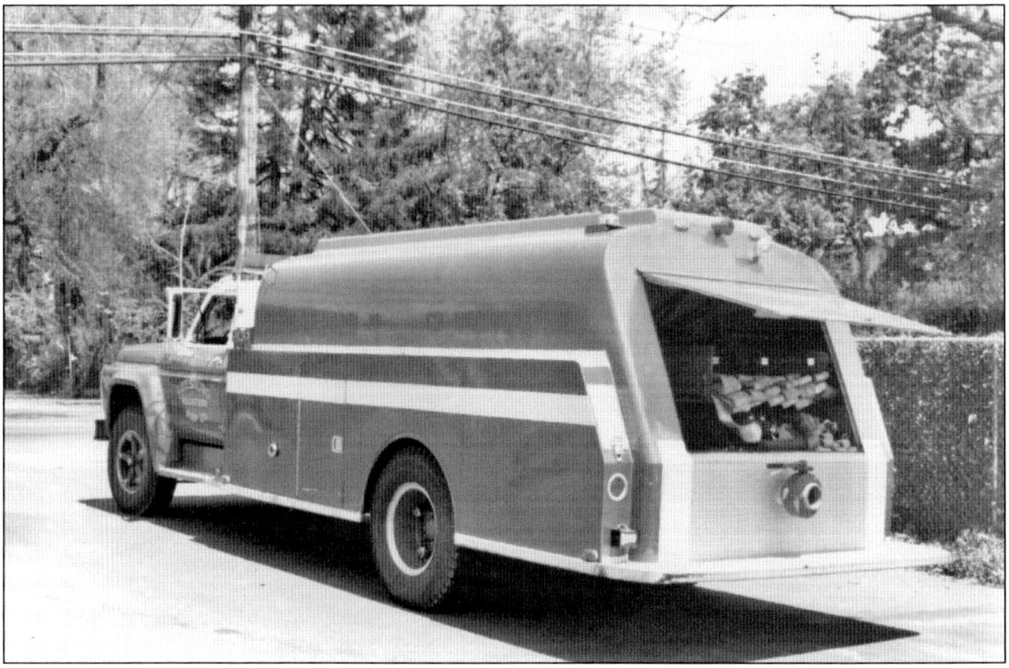

In 1975, Leesburg's new Tanker No. 1 was placed in service with this unusual Ford F-700 truck. The tanker body came from O. G. Hall Oil Company. An unusual feature of the tanker was the rear compartment, which stored a variety of firefighting equipment. The tanker was later sold to Loudoun County and used at the county landfill. (Courtesy of J. B. Anderson.)

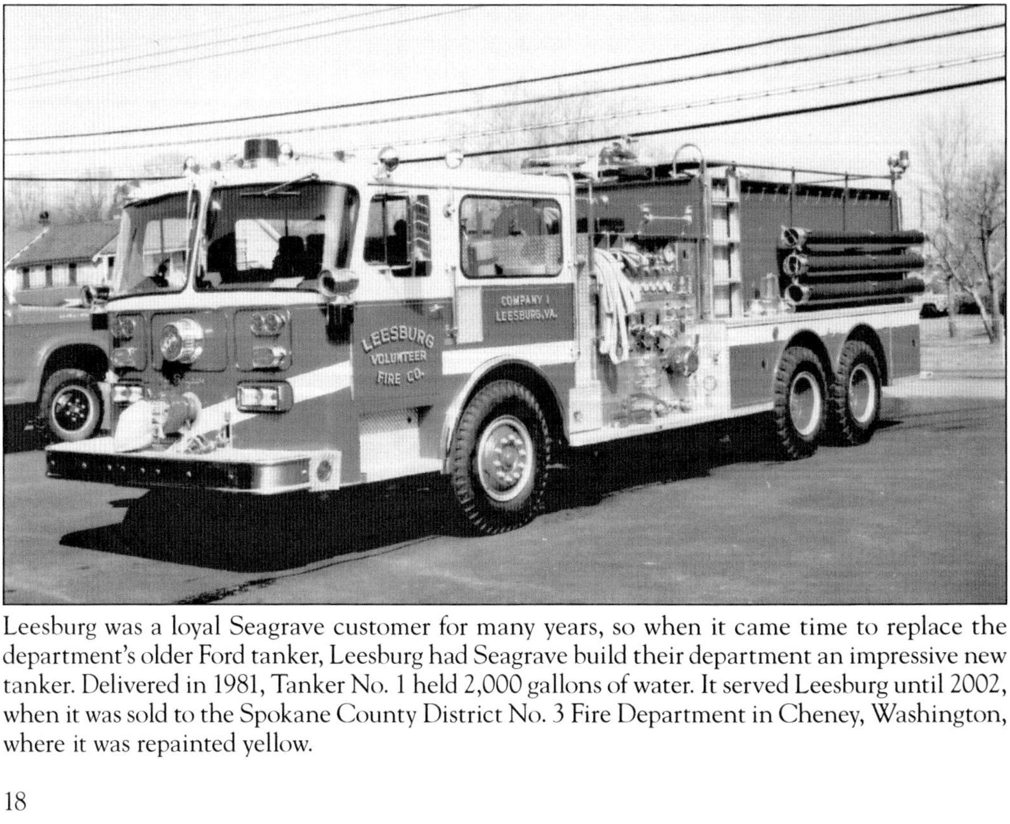

Leesburg was a loyal Seagrave customer for many years, so when it came time to replace the department's older Ford tanker, Leesburg had Seagrave build their department an impressive new tanker. Delivered in 1981, Tanker No. 1 held 2,000 gallons of water. It served Leesburg until 2002, when it was sold to the Spokane County District No. 3 Fire Department in Cheney, Washington, where it was repainted yellow.

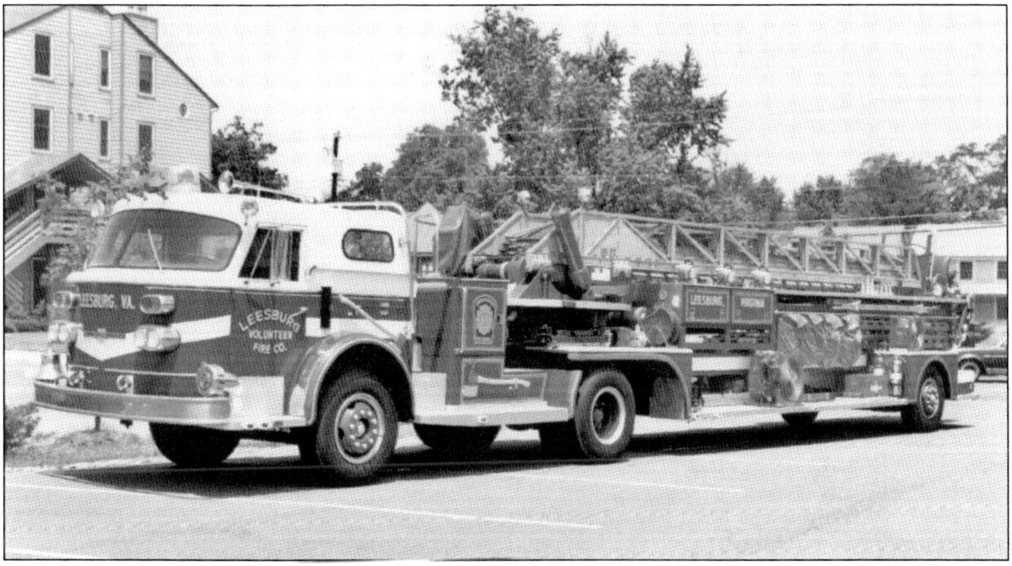

In 1987, Leesburg purchased its second tiller truck. Leesburg desired a backup to the 1973 Seagrave and was considering operating a second truck from the planned second station. This 1971 American LaFrance (ALF) 100-foot ladder truck was purchased from the Aetna Hose, Hook, and Ladder Company of Newark, Delaware, for $60,000. Originally an open-cab fire truck, it was fitted with a roof during the early 1980s. After this, the Leesburg apparatus roster sported two tiller trucks. Like the 1973 Seagrave tiller, the ALF was sold to Baltimore City, which affectionately nicknamed the ladder truck "Barney." After Baltimore, a collector from France purchased Barney. The ALF was put on a ship and made its journey to France, where it was photographed in front of the famous Eiffel Tower. (Below, courtesy of J. B. Anderson.)

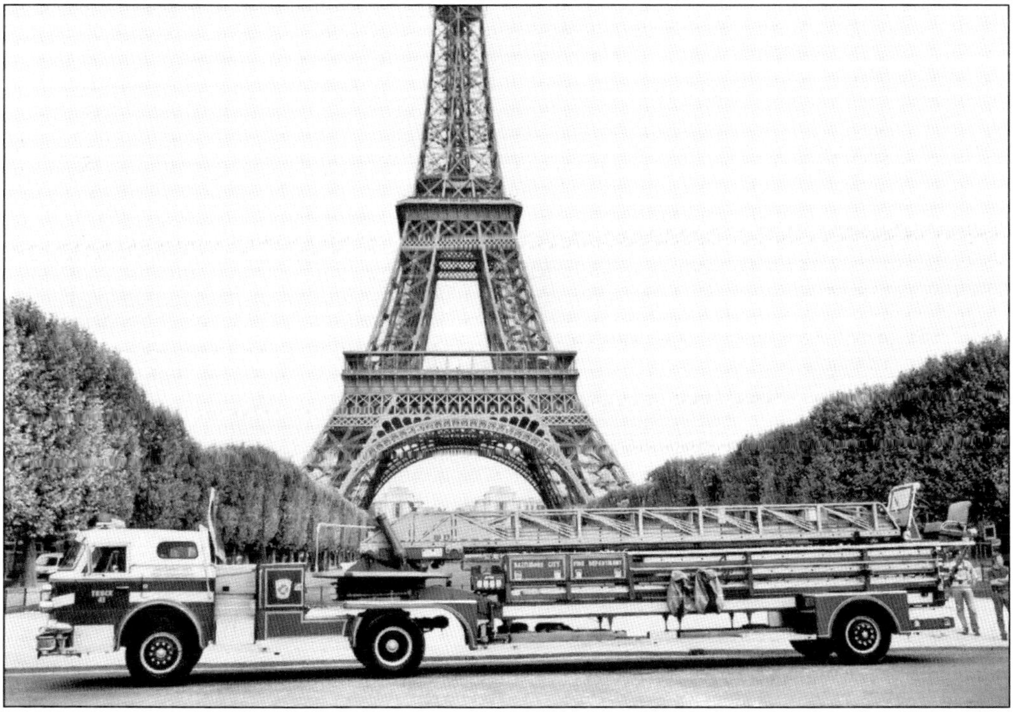

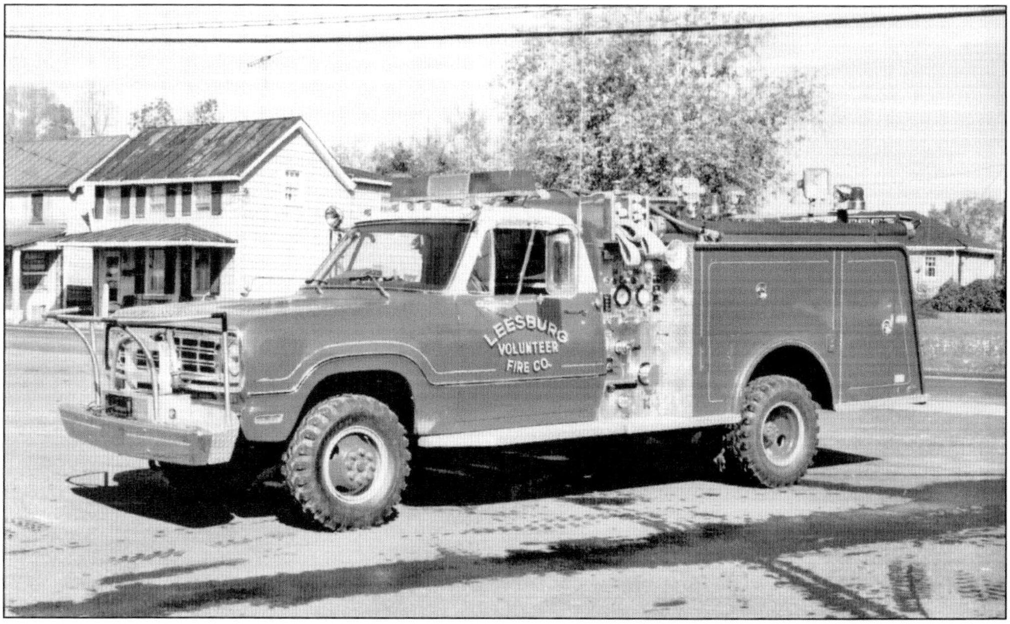

The mini-pumper concept became a popular choice for many departments. In 1976, Leesburg placed into service their new "Attack 1." Costing $30,000, it was a Dodge built by Hamerly, a fire-truck manufacturer. For brush fires, the front bumper could hold a firefighter who had access to a hose line. Heavily used by Leesburg, Attack 1 was sold to the Bladensburg, Maryland, VFD, located in Prince George's County. (Courtesy of Chuck Madderom.)

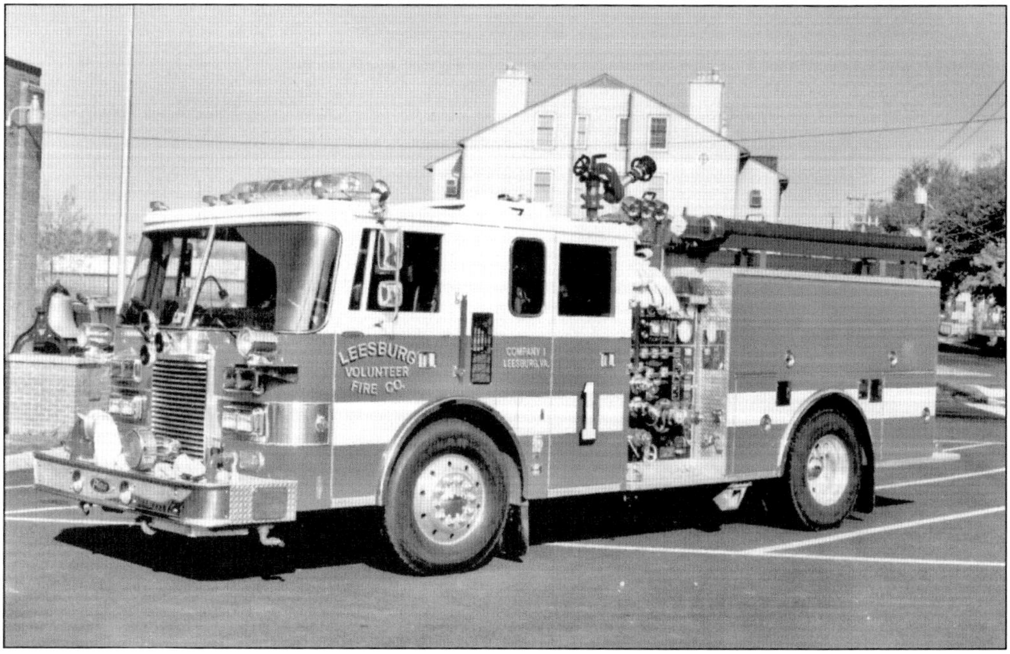

Leesburg's first four-door fire truck was the 1946 Buffalo engine. Forty-two years later, Leesburg placed into service their new Wagon No. 1, a four-door 1988 Pierce Lance. The Pierce replaced a wrecked 1982 Seagrave engine. Pierce, which manufactures fire trucks from Appleton, Wisconsin, equipped the engine with a 1,500-gallon-per-minute pump and a 750-gallon water tank.

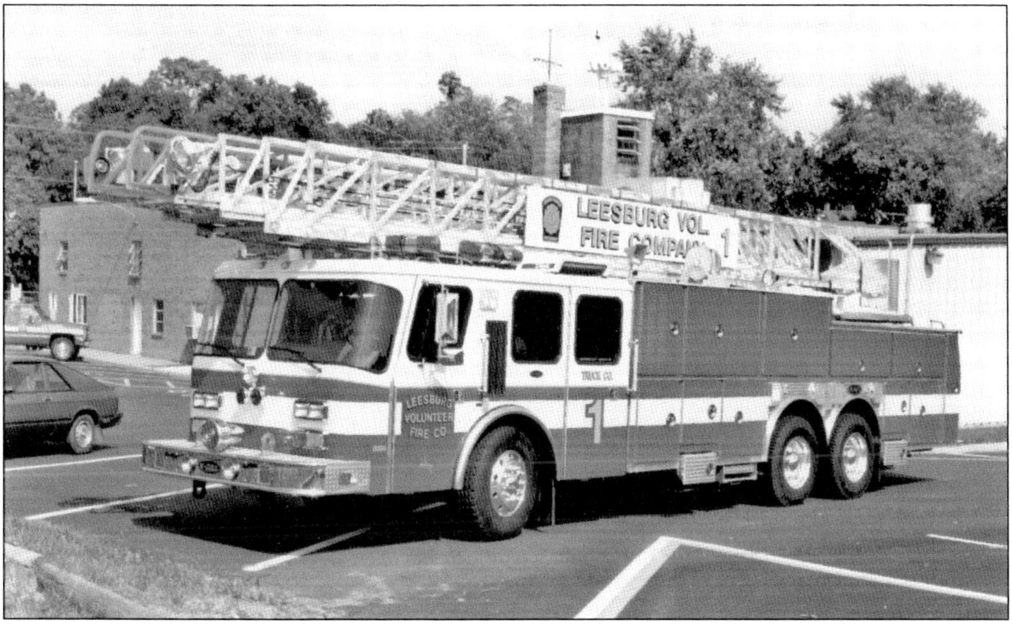

When Leesburg's 1992 E-One Hurricane 110-foot, rear-mount aerial was delivered, the Leesburg VFD then owned three aerial ladder trucks! The E-One joined the 1973 Seagrave and 1969 American LaFrance tiller trucks. Shortly afterward, both tillers were sold to Baltimore City, Maryland. After running as Truck No. 1 from the Loudoun Street station, the truck was later transferred to the new Station No. 20, located on Plaza Street.

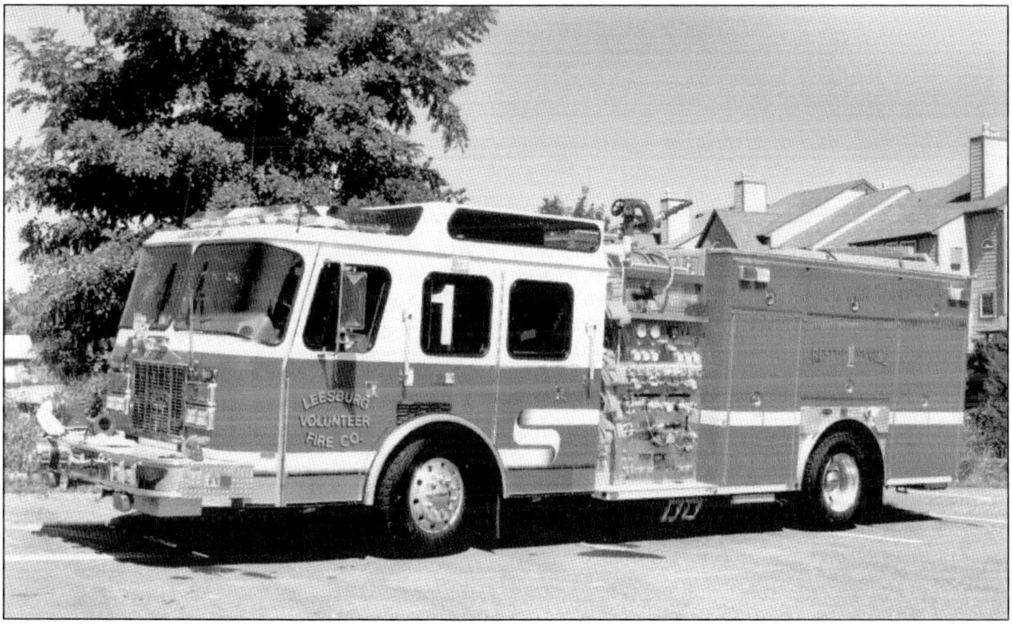

One of the first rescue engines in Northern Virginia was Leesburg's Rescue Engine No. 1. The concept of a rescue engine is to have a combination of a fire engine and a heavy rescue squad combined into one—one fire truck serving both purposes. Rescue Engine No. 1 carries typical firefighting equipment, as well as extrication tools, air bags, and other special tools. Rescue Engine No. 1 is a 1995 E-One Protector.

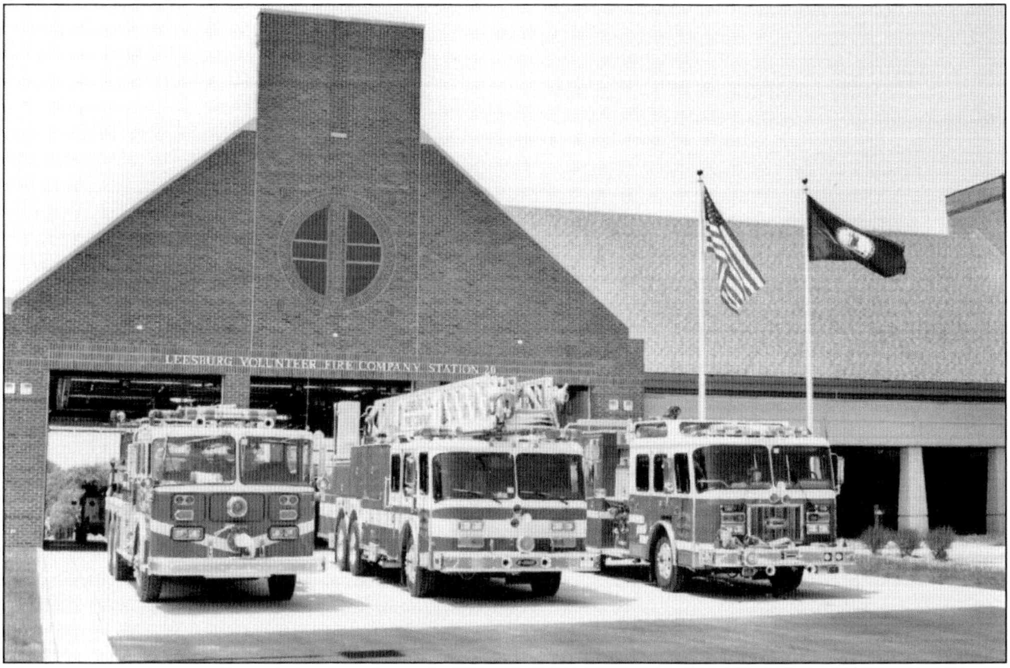

In 1998, Leesburg opened its second station on Plaza Street. Costing approximately $1.2 million, the station has three drive-through apparatus bays and space for training, bunkrooms, a kitchen, and a watch room, among other things. Photographed shortly after opening in 1998, Tanker No. 1, Truck No. 1, and Rescue Engine No. 1 sit on the front apron. Although designated as Station No. 20, all apparatus operating from the station have the unit designation of "1."

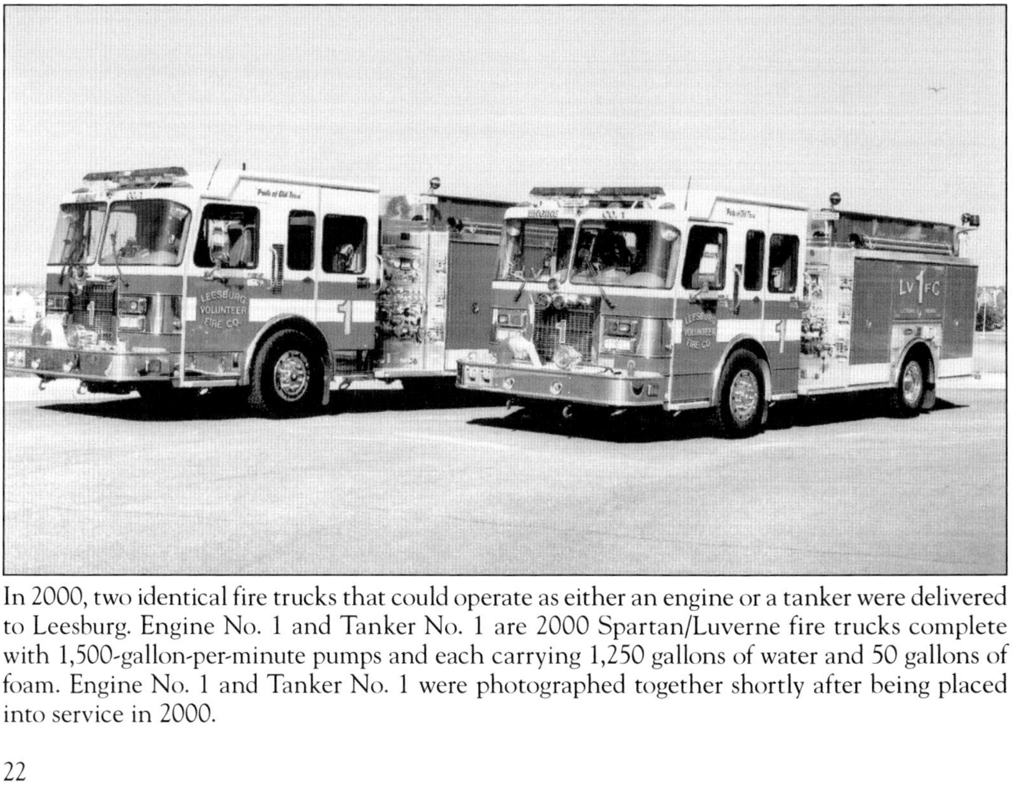

In 2000, two identical fire trucks that could operate as either an engine or a tanker were delivered to Leesburg. Engine No. 1 and Tanker No. 1 are 2000 Spartan/Luverne fire trucks complete with 1,500-gallon-per-minute pumps and each carrying 1,250 gallons of water and 50 gallons of foam. Engine No. 1 and Tanker No. 1 were photographed together shortly after being placed into service in 2000.

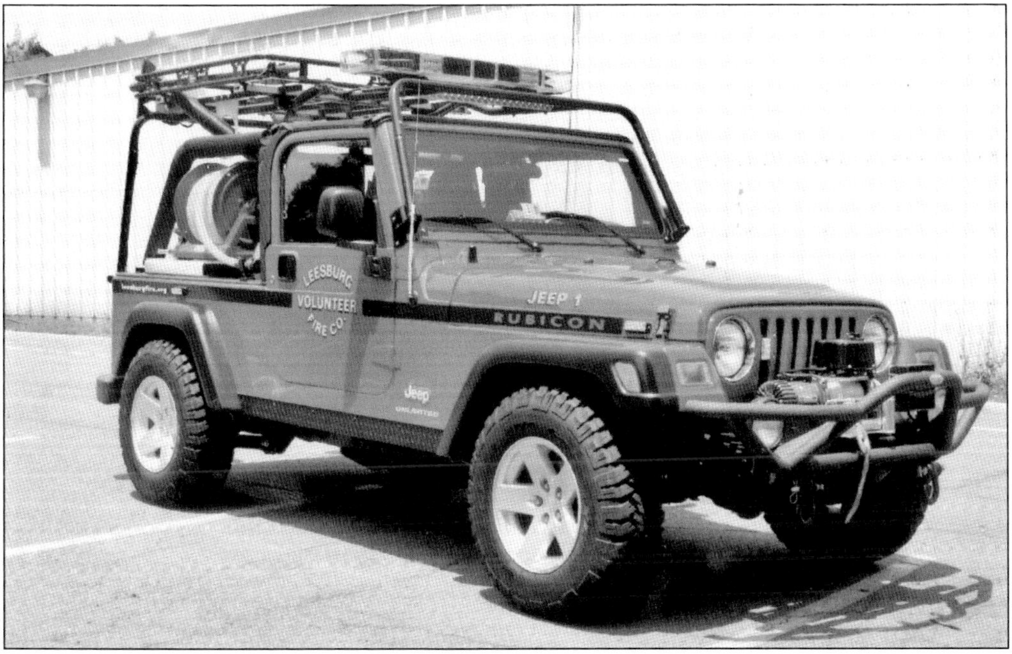

The Jeep tradition continued in Leesburg when the department took delivery of this 2006 model. With a radio designation of "Jeep 1," it is a 2006 Jeep Wrangler Rubicon that was outfitted by RKO Enterprises of Madison, Indiana, and includes a small 25-gallon-per-minute pump, a 60-gallon water tank, and a variety of tools and equipment to handle outdoor fires.

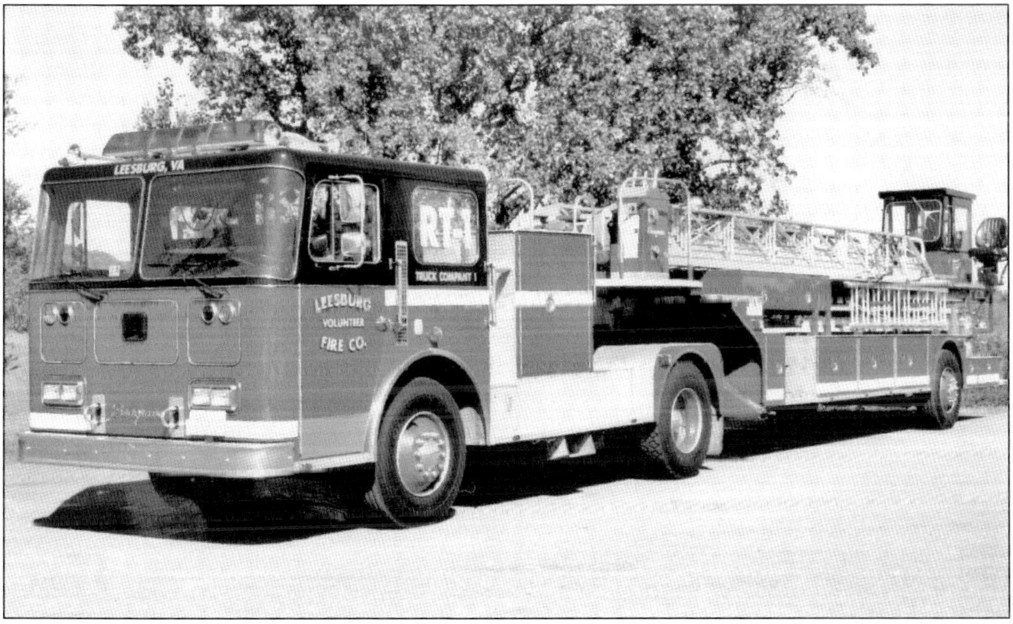

The tiller truck returned to Leesburg in 2006, when the department purchased a used 1988 Seagrave 100-foot ladder truck. In preparation for the delivery of a new Seagrave ladder truck in 2007, Leesburg purchased the 1988 model to be used primarily for driver training. Reserve Truck No. 1 is also painted in the department's new color of black over red. The Seagrave originally saw service in Richmond, Virginia, and later in Washington, D.C.

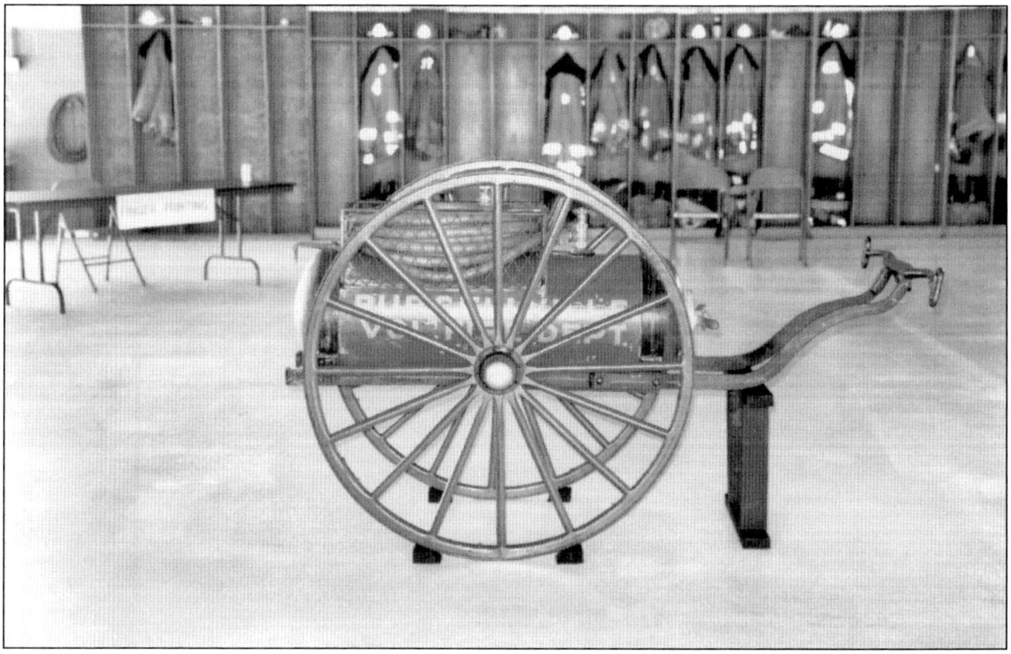

Purcellville, in the early 1900s, had several large fires. The town of Purcellville acquired two hand-drawn chemical wagons, and the residents of the community used them to help fight fire. This hand-drawn wagon was built in 1915 and is still owned by the Purcellville Volunteer Fire Department. (Courtesy of Purcellville VFD.)

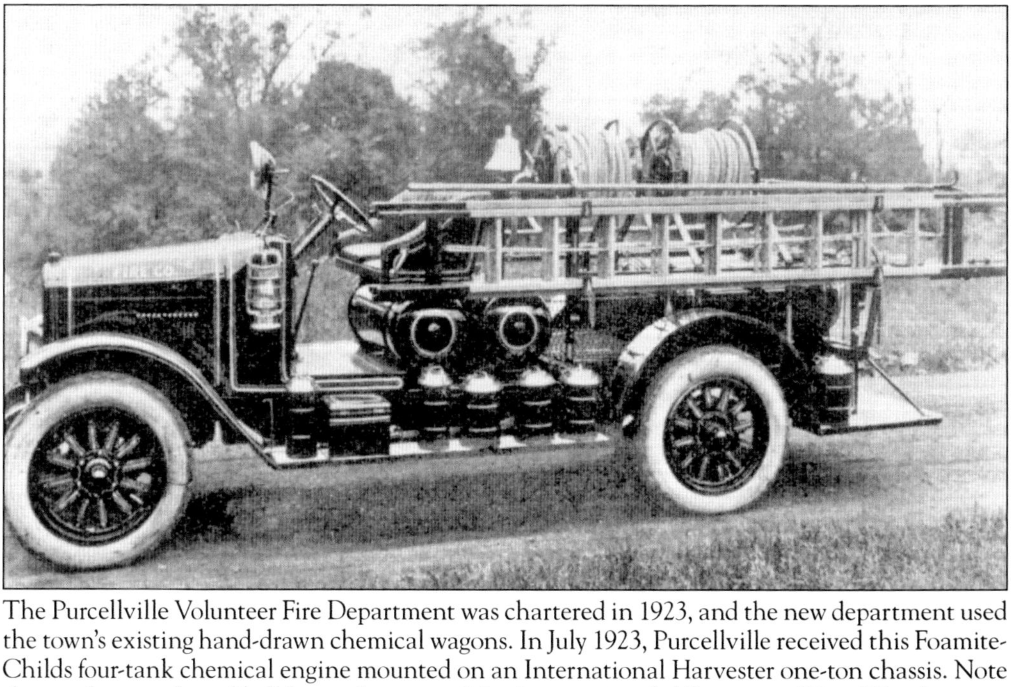

The Purcellville Volunteer Fire Department was chartered in 1923, and the new department used the town's existing hand-drawn chemical wagons. In July 1923, Purcellville received this Foamite-Childs four-tank chemical engine mounted on an International Harvester one-ton chassis. Note the two hose reels and bell located on top of the International. (Courtesy of Pete West.)

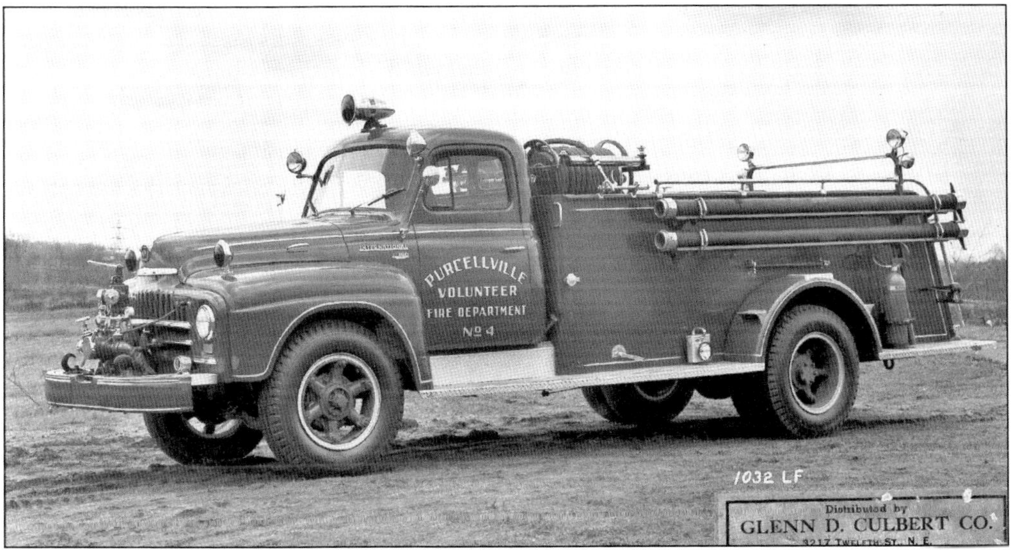

This delivery photograph shows Purcellville's 500-gallon-per-minute, front-mount pump 1949 International built by the American Fire Apparatus Company. Nicknamed "Old Binder," it was initially painted red and later painted lime yellow and white. Note the No. 4 on the door. This represented the fourth fire truck owned and operated by Purcellville. The International was sold to a farm in the Purcellville area.

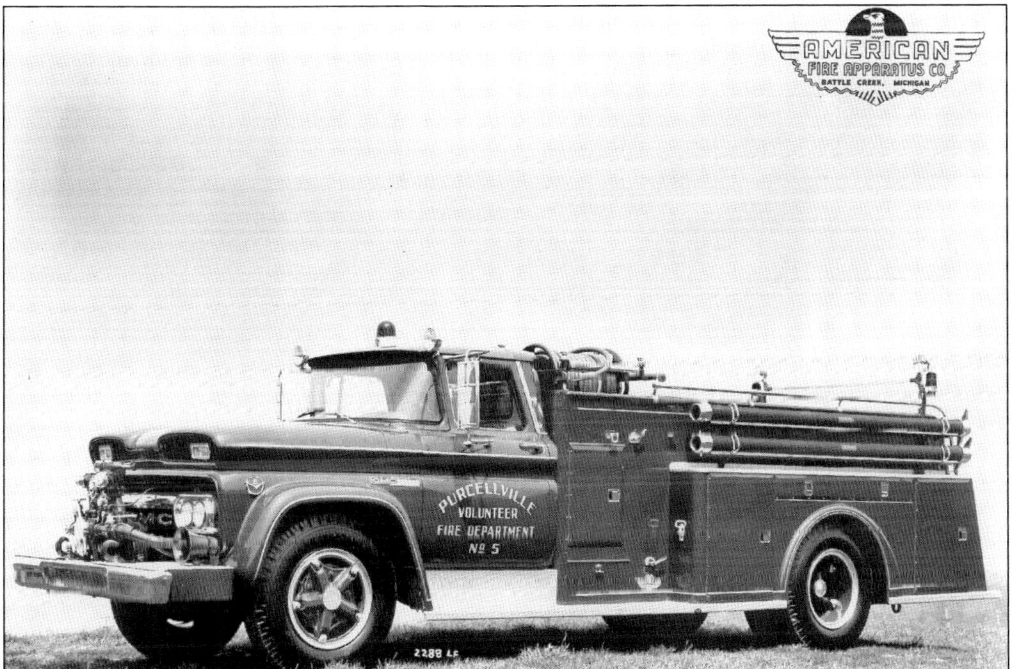

Purcellville's striking 1960 GMC/American was equipped with a 500-gallon-per-minute, front-mount pump and held 500 gallons of water. Like many fire companies in the Northern Virginia area, Purcellville's department had Glenn D. Culbert Company handle the purchase of this truck. This 1960 GMC remained the color red and was never repainted to Purcellville's newer colors of lime yellow and white.

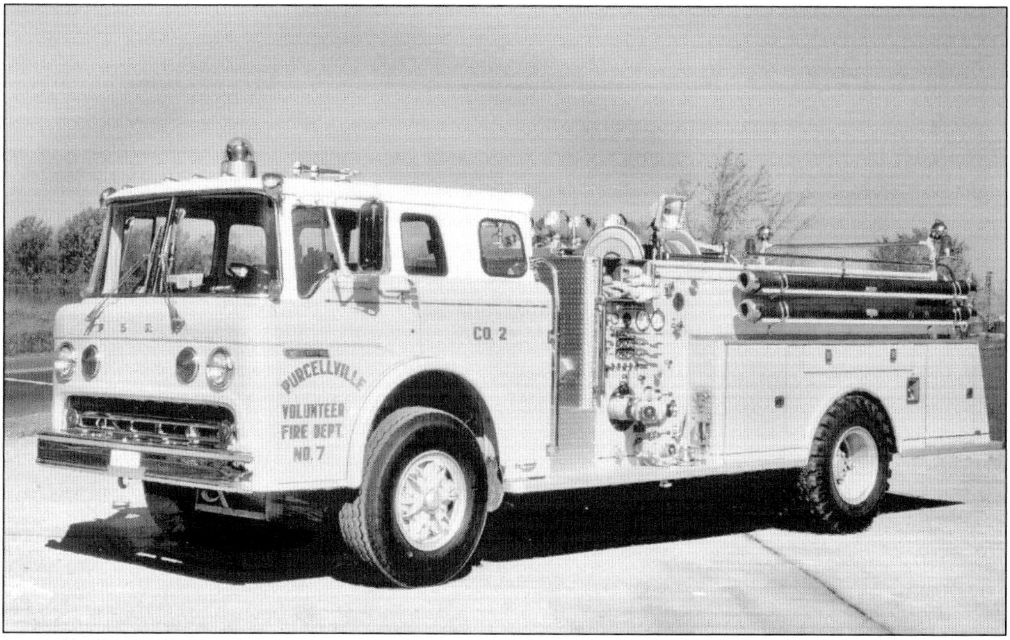

Purcellville made the decision to change their apparatus color from red to lime yellow and white in the mid-1970s. The first fire truck to arrive in the new colors was this 1976 Ford C built by American. Equipped with a 750-gallon-per-minute pump and carrying 600 gallons of water, it was later sold to the Shenandoah Shores, Virginia, Volunteer Fire Department. (Courtesy of Robert Kimball.)

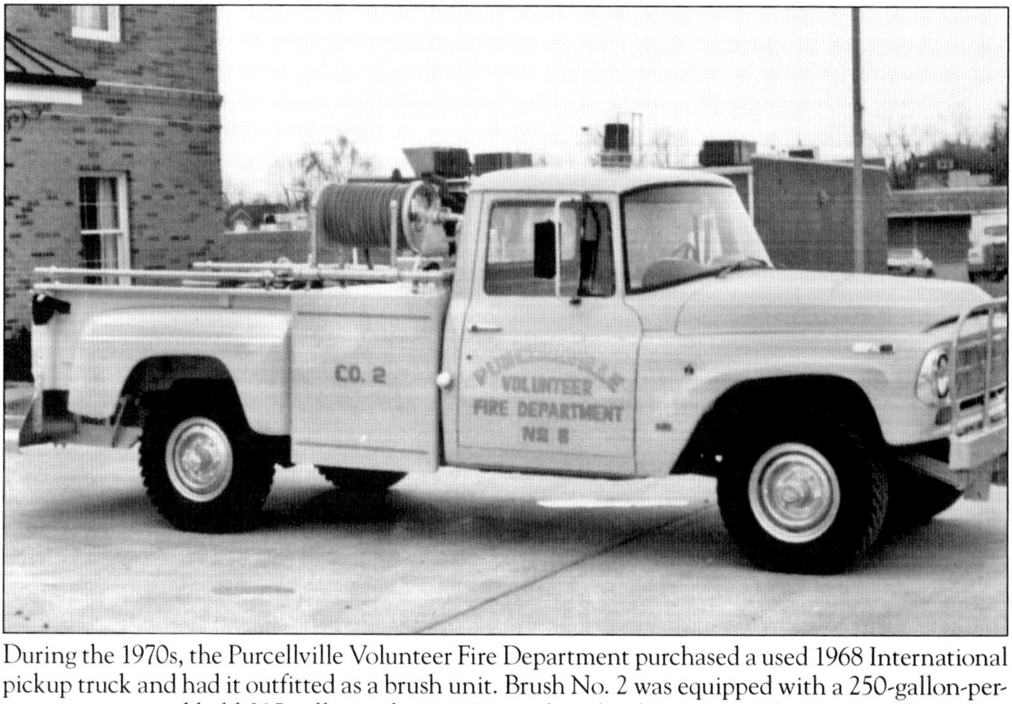

During the 1970s, the Purcellville Volunteer Fire Department purchased a used 1968 International pickup truck and had it outfitted as a brush unit. Brush No. 2 was equipped with a 250-gallon-per-minute pump and held 385 gallons of water. Painted in the department's lime yellow and white colors, it was replaced in 1980 with a Chevrolet pickup truck. (Courtesy of Purcellville VFD.)

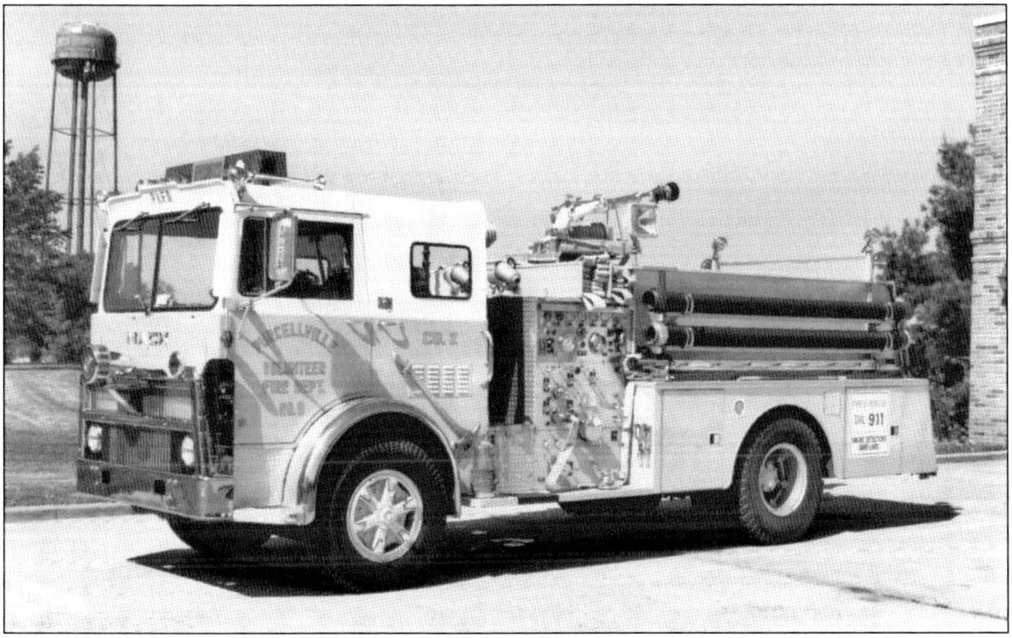

Prior to Purcellville purchasing this 1979 Mack MB/Hamerly, it was used by Hamerly as a demonstrator fire truck. The Mack was equipped with a 1,000-gallon-per-minute pump. The large water-deck gun located on top of the Mack was acquired from the Arlington County, Virginia, fire department. Like some other apparatus in Loudoun County, the Mack was sold to Tazewell County, Virginia.

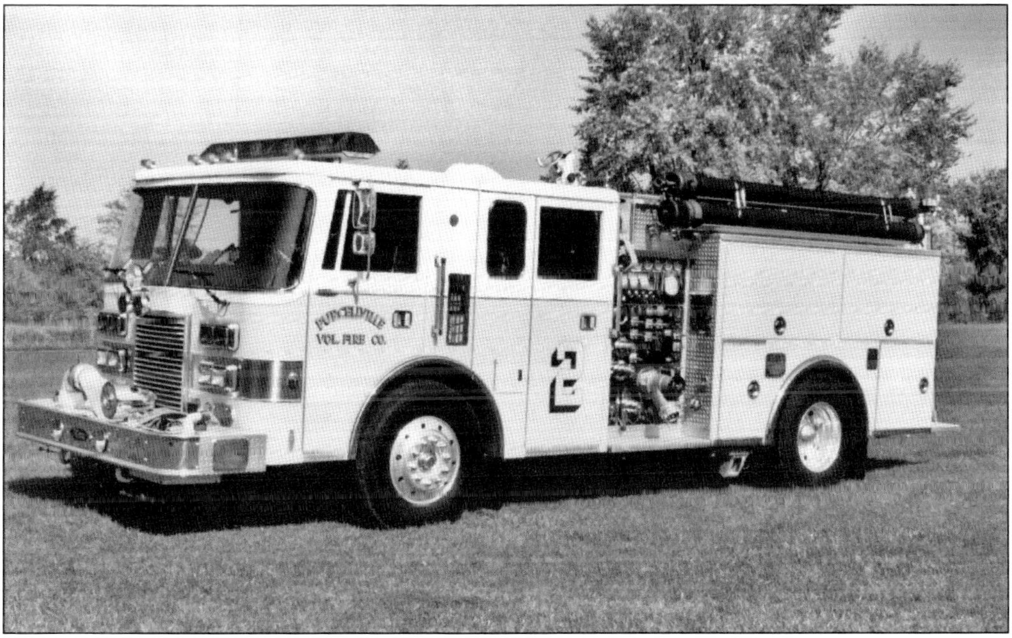

Purcellville's first four-door enclosed fire truck was this 1988 Pierce Lance equipped with a 1,250-gallon-per-minute pump and carrying 750 gallons of water. It has the radio designation of Wagon 2. The Pierce was photographed in the grass next to the current Station No. 2 on North Maple Avenue.

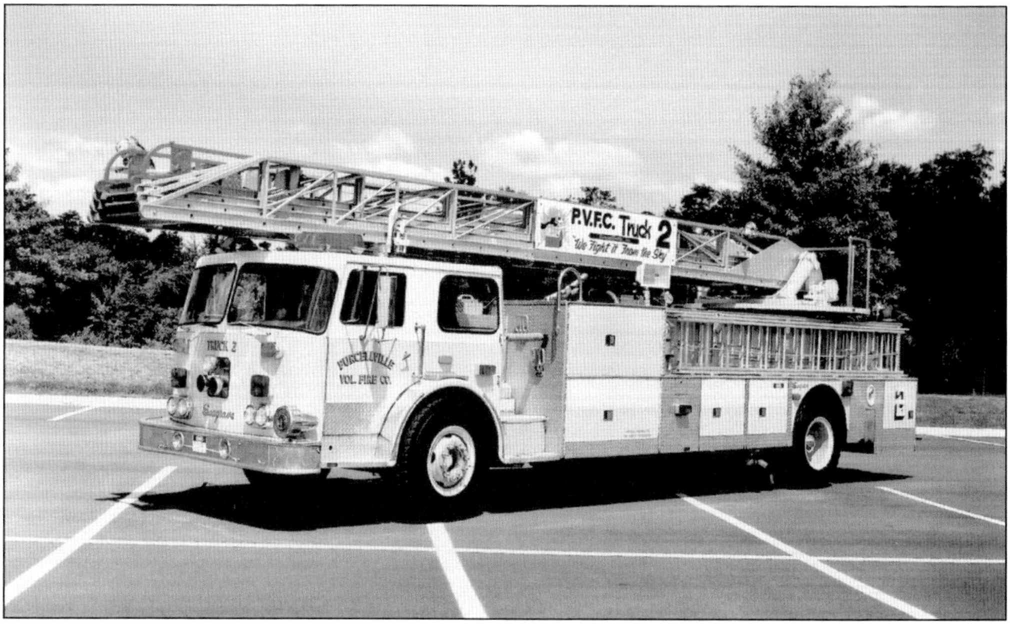

Truck No. 2 from Purcellville was purchased from the Sterling Volunteer Fire Department in 1991. After acquiring the 1974 Seagrave 100-foot, rear-mount aerial, Purcellville had it sent to the Mecklenburg, Virginia, Department of Corrections, where the inmates reconditioned it. The Seagrave ladder, with a ladder sign that says, "We Fight 'it' From the Sky," was sold to the Circleville, West Virginia, Volunteer Fire Department.

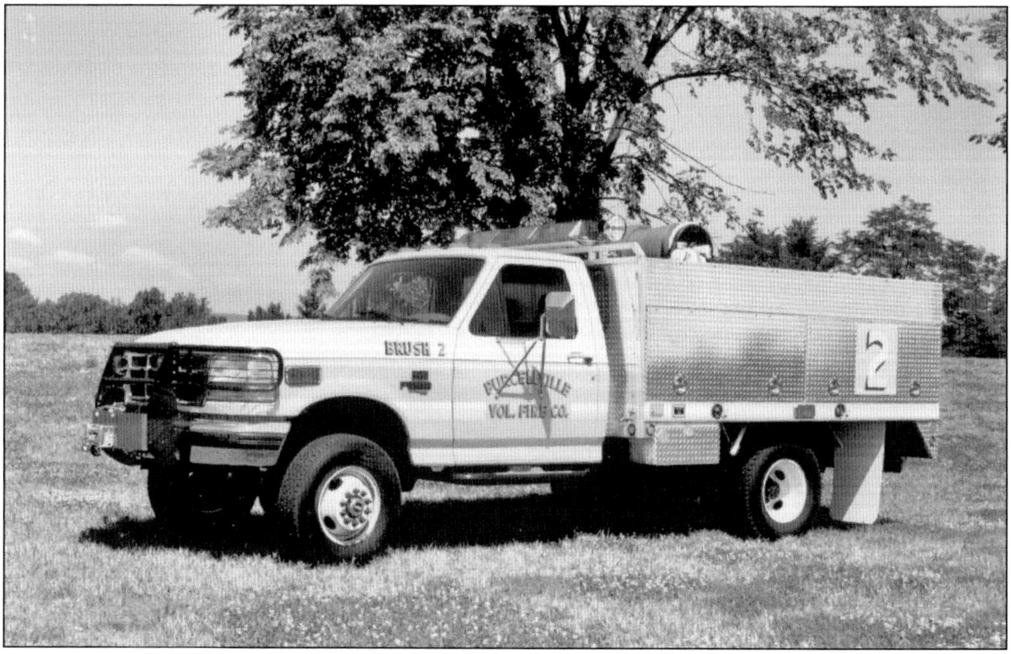

Purcellville's Brush No. 2 is equipped to operate as a brush truck as well as to assist in water-supply operations. In 1997, Brush No. 2 received this Ford F Super Duty, which replaced a 1980 Chevrolet brush truck. With its 500-gallon-per-minute pump and smaller size, Brush 2 can maneuver better in tight areas than standard-size fire trucks. It is also equipped with a 300-gallon water tank.

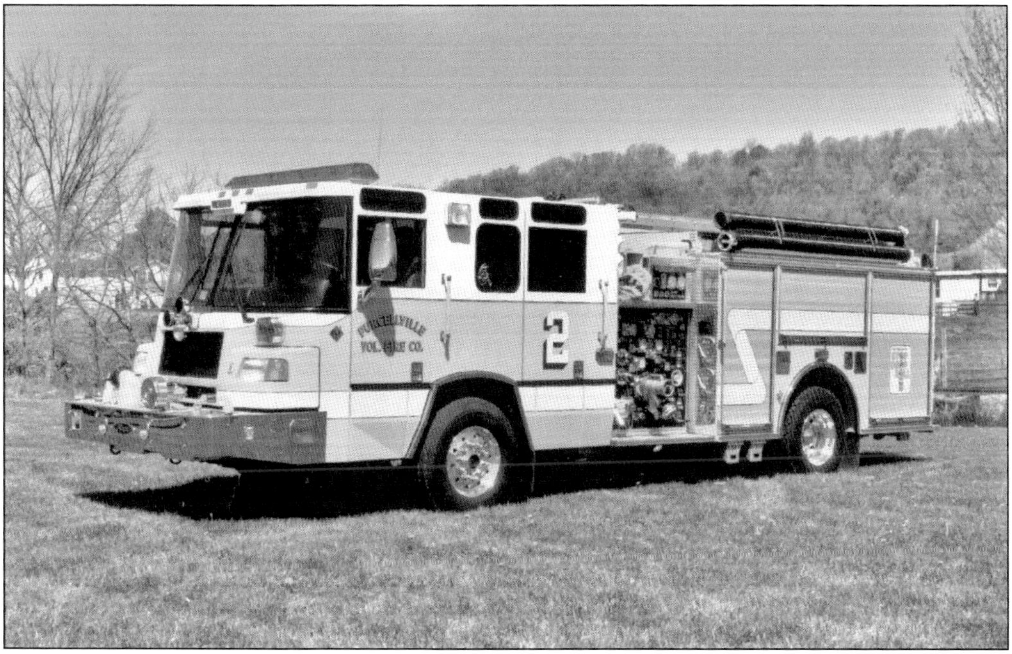

Brand-new and posed for a picture, Purcellville's Engine No. 2 was photographed at the fireman's parade at the Winchester, Virginia, Apple Blossom Festival in 1999. Engine No. 2 is a 1999 Pierce Quantum equipped with a 1,250-gallon-per-minute pump, 750 gallons of water, and roll-up compartment doors.

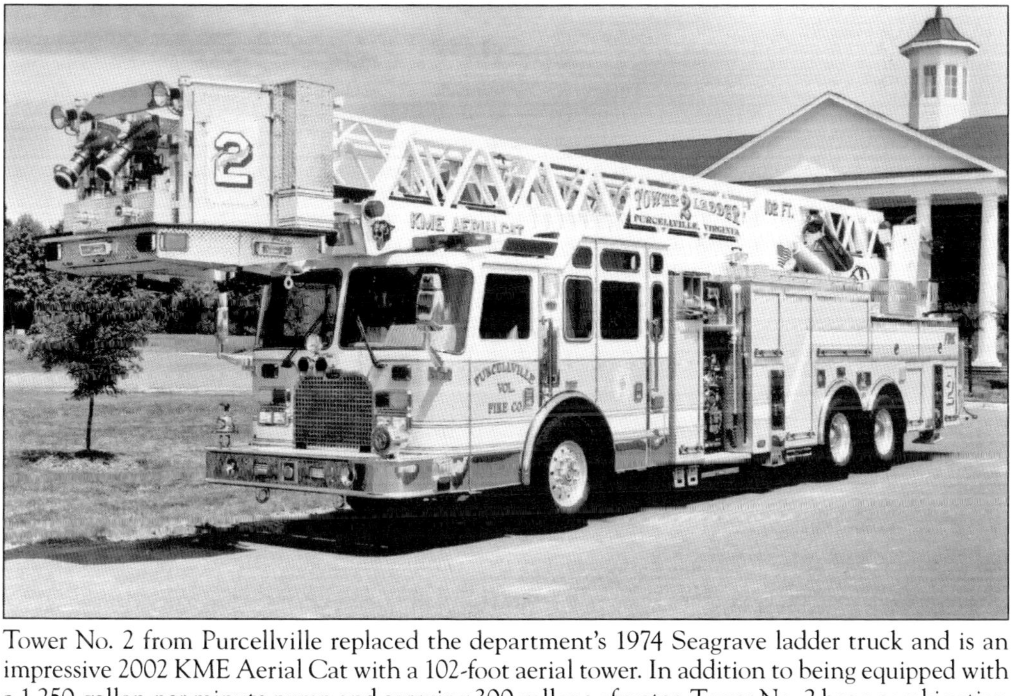

Tower No. 2 from Purcellville replaced the department's 1974 Seagrave ladder truck and is an impressive 2002 KME Aerial Cat with a 102-foot aerial tower. In addition to being equipped with a 1,250-gallon-per-minute pump and carrying 300 gallons of water, Tower No. 2 has a combination of regular compartment doors as well as roll-up doors. The tower was photographed in 2003 in front of Patrick Henry College, located in Purcellville.

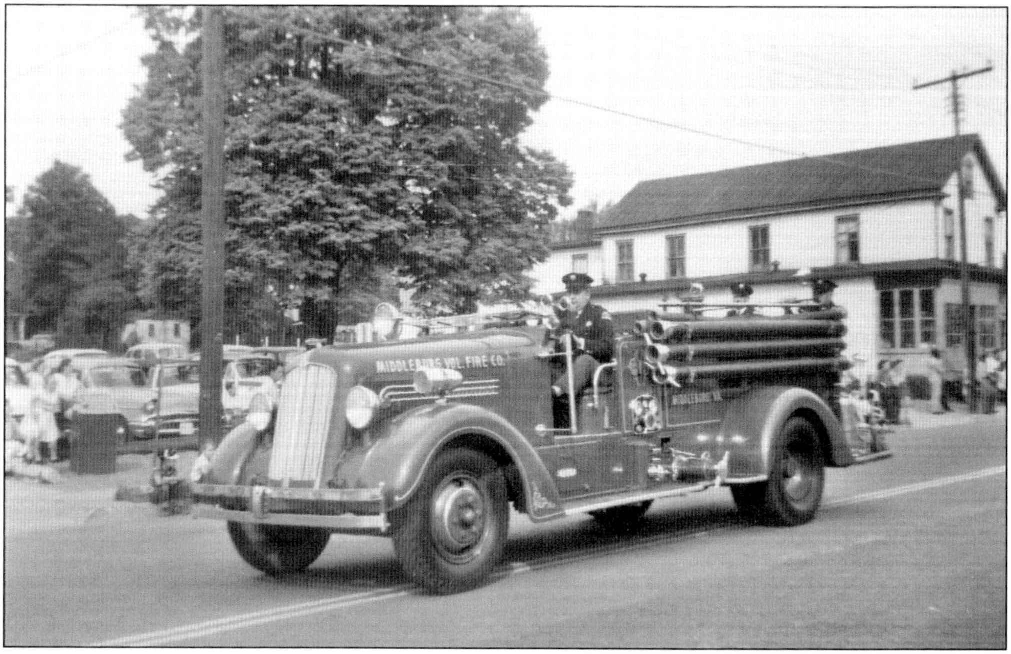

In 1936, the Middleburg Volunteer Fire Department was organized as Station No. 3 in Loudoun County. The department's first fire truck was this 1936 Seagrave, which cost $6,000. Seagrave also donated six coats and six pairs of boots to help the young department. The Seagrave had a front windshield that folded down, as can be seen in this photograph. (Courtesy of Pete West.)

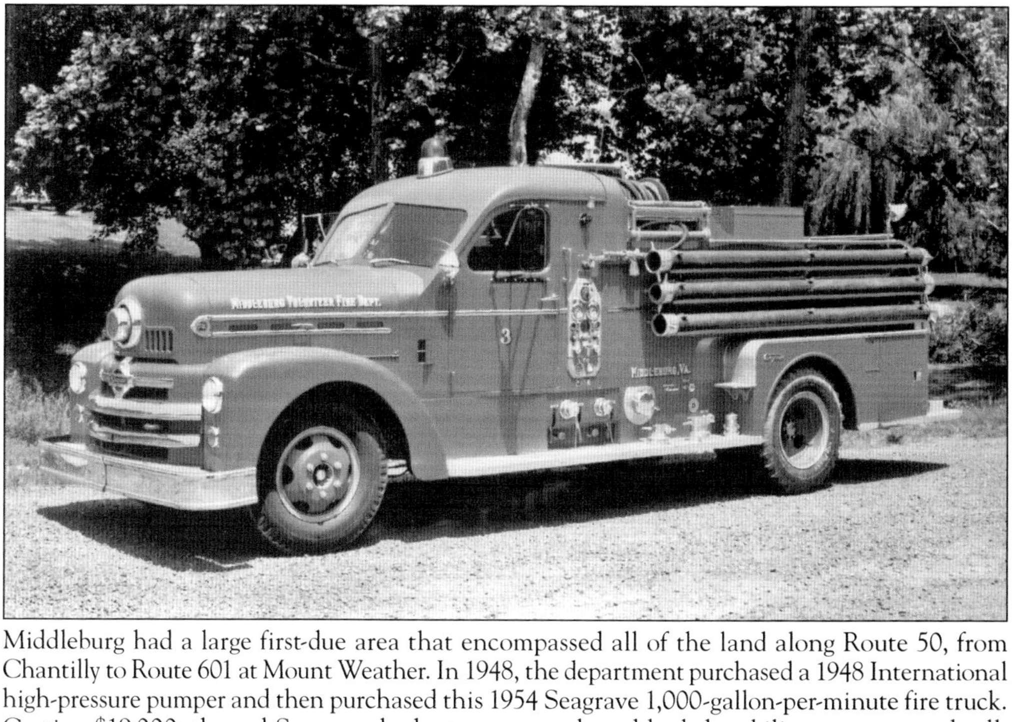

Middleburg had a large first-due area that encompassed all of the land along Route 50, from Chantilly to Route 601 at Mount Weather. In 1948, the department purchased a 1948 International high-pressure pumper and then purchased this 1954 Seagrave 1,000-gallon-per-minute fire truck. Costing $19,000, the red Seagrave had a two-man cab and had the ability to pump and roll. (Courtesy of Chuck Madderom.)

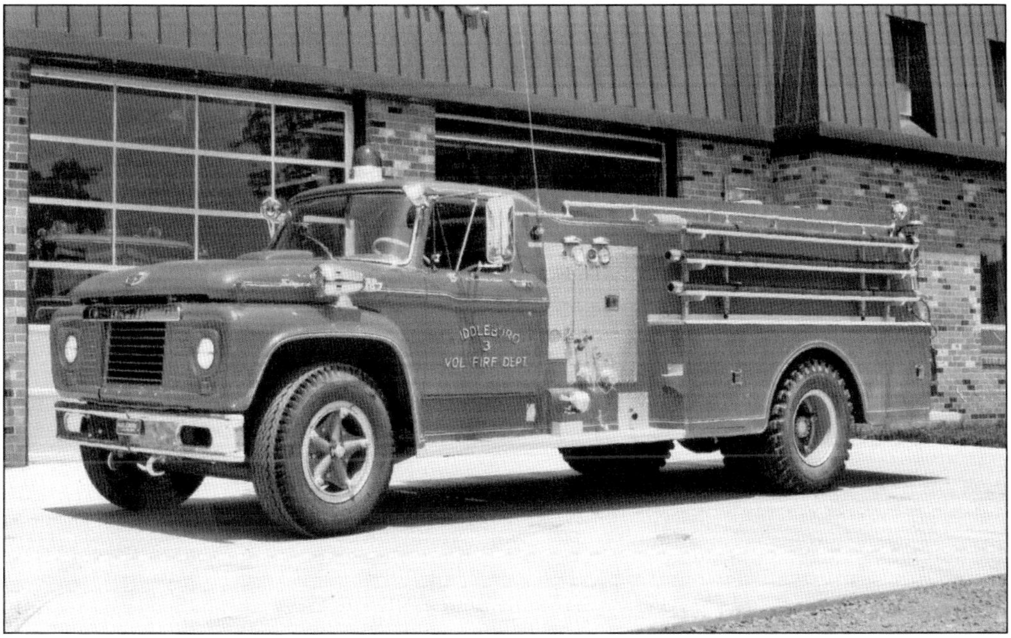

Tanker No. 3 from Middleburg was this red 1965 Ford F with a fire-truck body built by Howe. It was equipped with a 1,500-gallon water tank. During the 1970s, Middleburg changed their apparatus color from red (or red and white) to yellow. This 1965 tanker was later repainted to yellow and was eventually sold to the Blue Ridge Volunteer Fire Department in nearby Clarke County. (Courtesy of Chuck Madderom.)

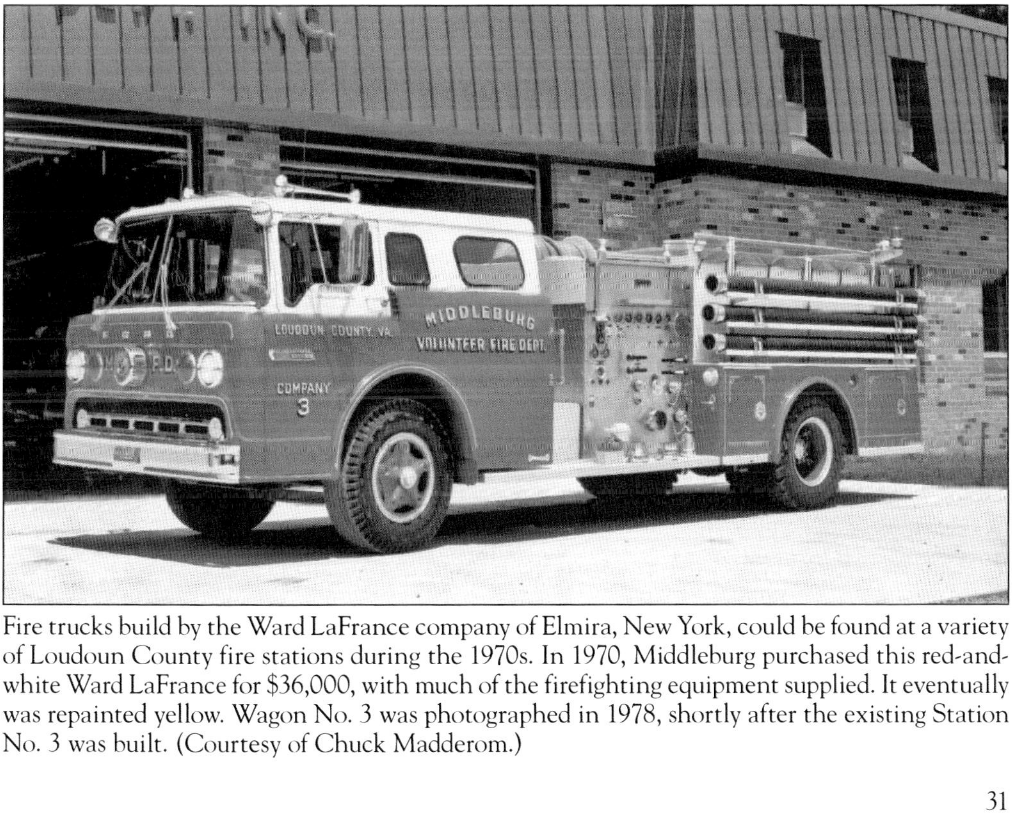

Fire trucks build by the Ward LaFrance company of Elmira, New York, could be found at a variety of Loudoun County fire stations during the 1970s. In 1970, Middleburg purchased this red-and-white Ward LaFrance for $36,000, with much of the firefighting equipment supplied. It eventually was repainted yellow. Wagon No. 3 was photographed in 1978, shortly after the existing Station No. 3 was built. (Courtesy of Chuck Madderom.)

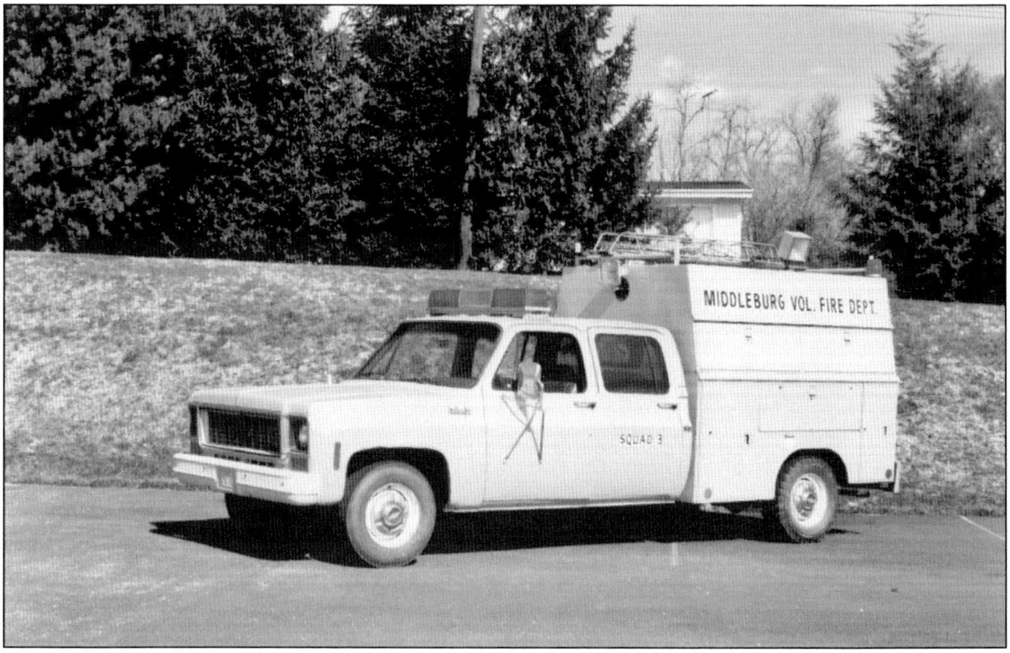

During the 1970s, Middleburg operated a small equipment truck. It was a 1960s Dodge van donated to the department by C&P Telephone Company. The department replaced the Dodge van when they purchased a used 1970 Chevrolet and had a Watson squad body added to it. Complete with four doors, it was painted yellow. (Courtesy of Dave Bowen.)

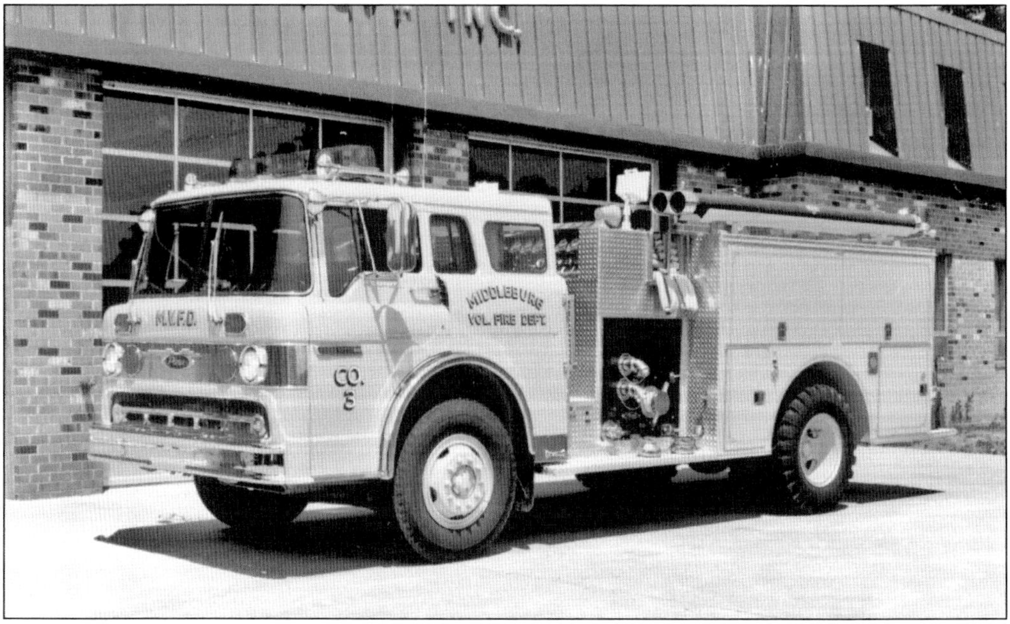

Many Middleburg members liked the smaller size of the new Wagon No. 3 when it was purchased in 1980. Built by Pierce on a Ford C chassis, Wagon No. 3 had a 1,000-gallon-per-minute pump and a top-mount pump panel. With a purchase price of approximately $80,000, this Ford was the department's first Pierce fire truck and the first new truck painted yellow. It was later sold to a fire department in Alabama.

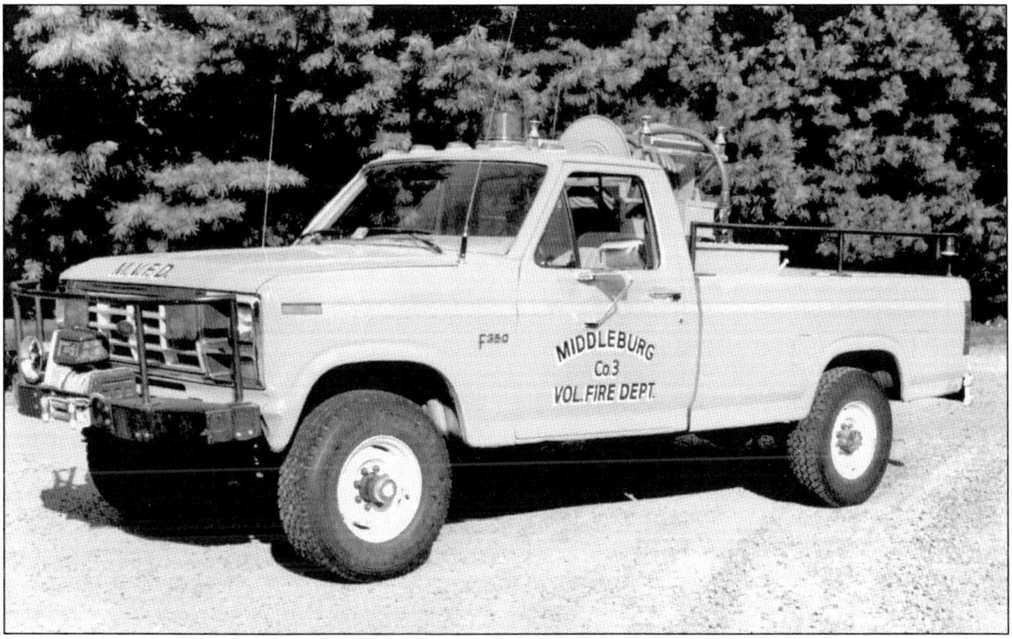

When Middleburg received their new brush truck in 1985, it arrived in a slightly different color than the yellow of the Middleburg apparatus. Brush No. 3 was delivered in a highway yellow (which was more orange), common to the Virginia Department of Transportation (VDOT) trucks. It was kept that color, and it replaced an older Jeep brush unit that Middleburg had received from Leesburg, Virginia.

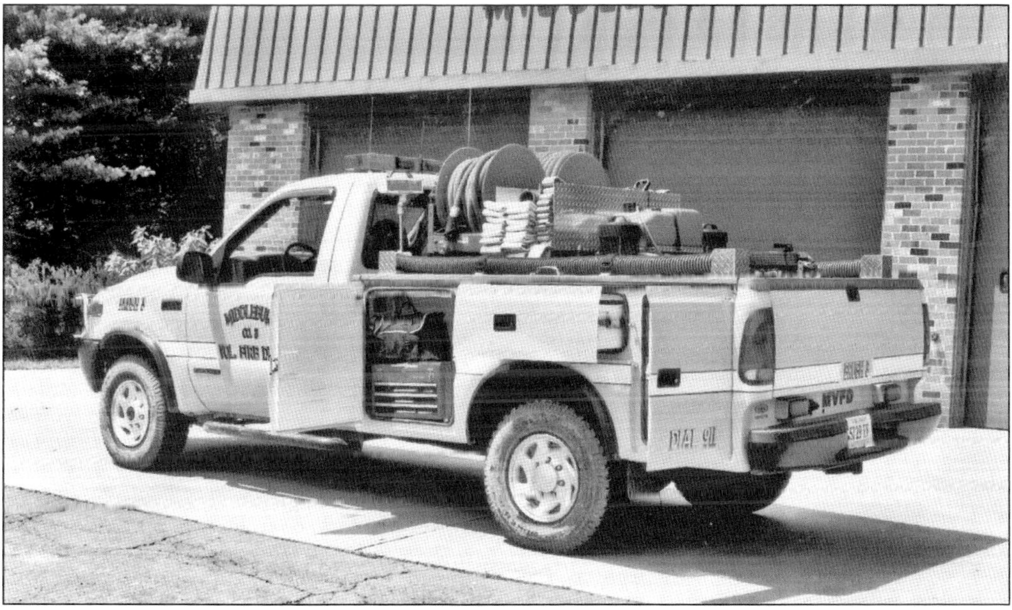

The 1985 "orange" brush truck from Middleburg was replaced in 2001 with this Ford. Brush No. 3 has unusual compartments built into the side body for storage of firefighting equipment. In addition to the dual-booster reels and other hose on top of the truck, Brush No. 3 carries 225 gallons of water. With mud on its tires after recently returning from a brush fire, Brush No. 3 was photographed in 2003.

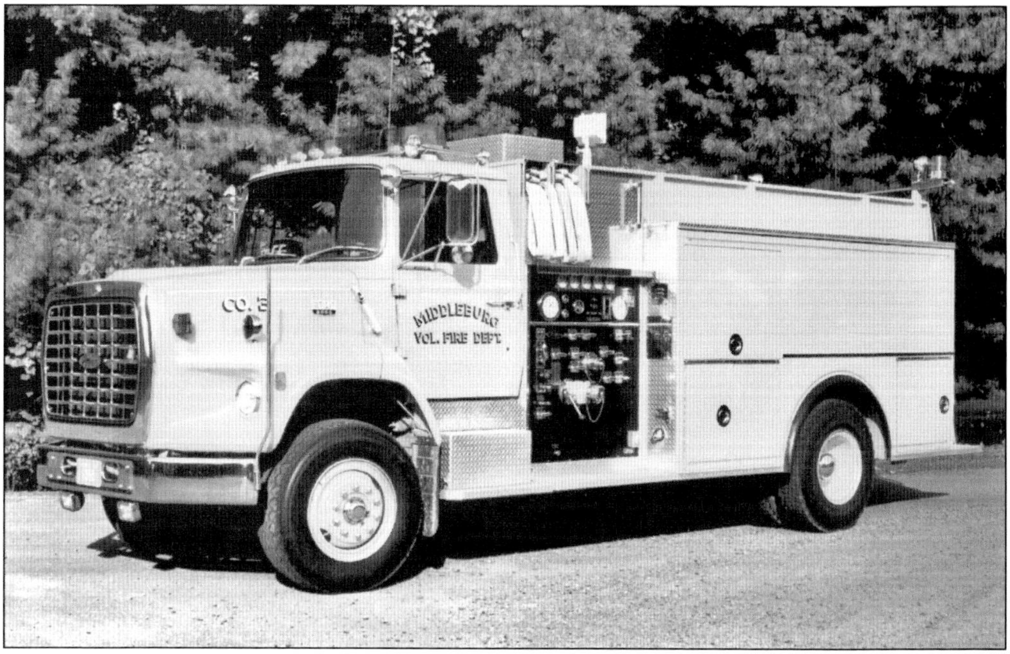

A second Pierce fire truck was delivered to Middleburg in 1986 and ran as Tanker No. 3. The 1986 Ford L8000 was equipped with a smaller, 400-gallon-per-minute pump and carried 1,500 gallons of water. Tanker No. 3 served the Middleburg community for over 20 years, and it was replaced in 2007.

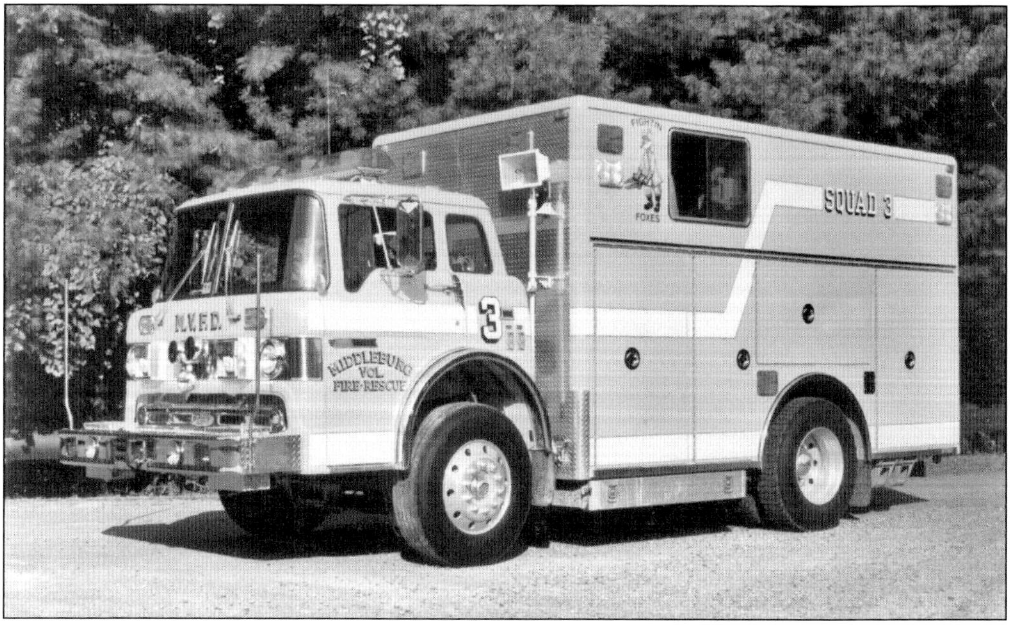

Middleburg continued purchasing commercial cab fire trucks and preferred trucks built by Pierce. The smaller 1970 Chevrolet/Watson squad was replaced in 1990 when Middleburg received this Pierce rescue squad built on a Ford C cab. Because one of the primary functions of Squad No. 3 was automobile extrication, the "Fightin' Foxes" mural on the squad showed a fox dressed in firefighting gear and holding an extrication tool.

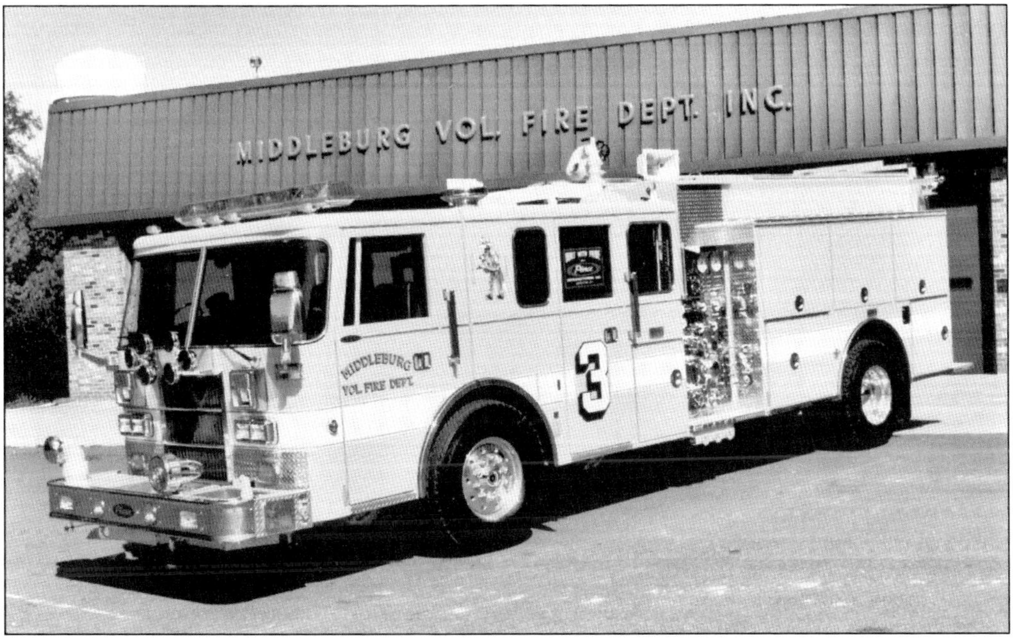

This 1992 Pierce Lance was Middleburg's first four-door fire truck, and it is equipped with a 1,250-gallon-per-minute pump and holds 1,000 gallons of water. A unique feature of Wagon No. 3 is the dual Roto-Ray warning lights that are mounted on the front of the cab. Wagon No. 3 was photographed in front of Station No. 3 shortly after being delivered. (Courtesy of Purcellville VFD.)

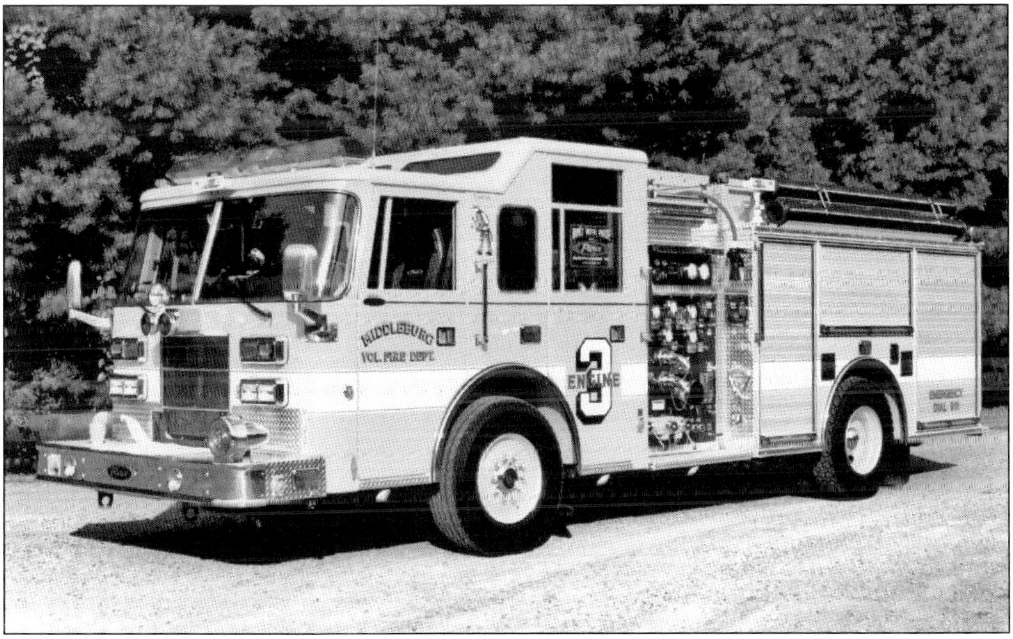

Middleburg returned to Pierce for another pumper when they ordered a fire truck in 1998. Middleburg specified roll-up compartment doors in addition to the 1,250-gallon-per-minute pump and 1,000-gallon water tank on this truck, built on a Pierce Saber model. The new Pierce Saber received the radio designation of "Engine 3."

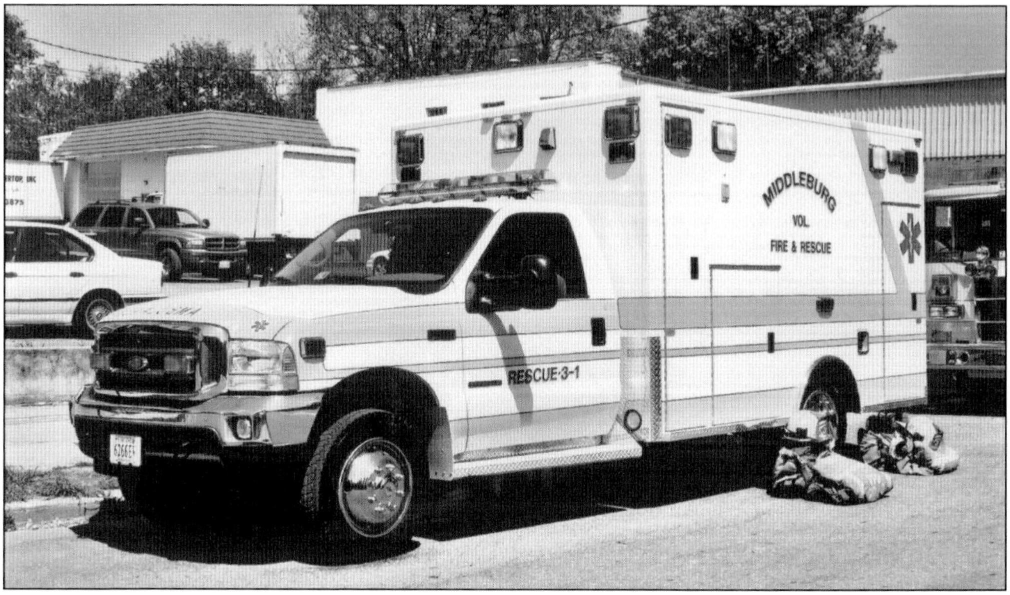

A local Ford dealer donated to Middleburg a 1953 Ford station wagon. It became Middleburg's first ambulance. Since that time, the department has used a variety of Fords as ambulances. In 2002, the department took delivery of this new Ford F-450 built by PL Custom Emergency Vehicles. The new Rescue 3-1 was displayed at the 2002 apple blossom firefighters' parade in Winchester, Virginia.

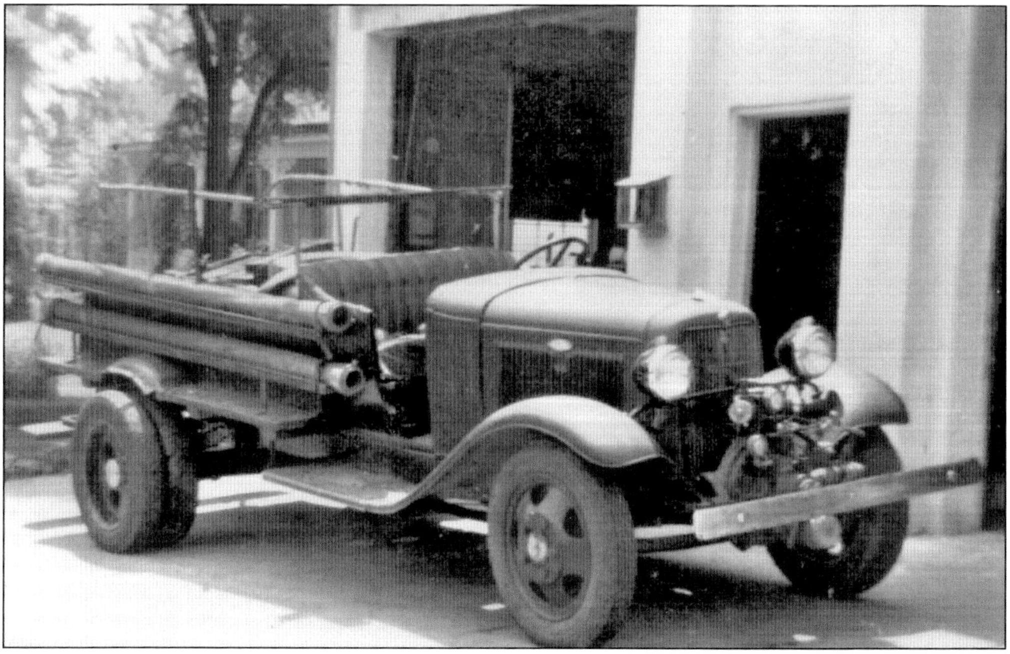

In the early 1900s, the Round Hill community had a hand-pulled hose wagon, which was eventually pulled by a Ford Model T. The fire department was organized in 1938 but was not officially chartered until 1948. The department's first fire truck was this 1934 Ford. It was purchased by a local resident and donated to Station No. 4. It was later sold to Amissville, Virginia. (Courtesy of Round Hill VFD.)

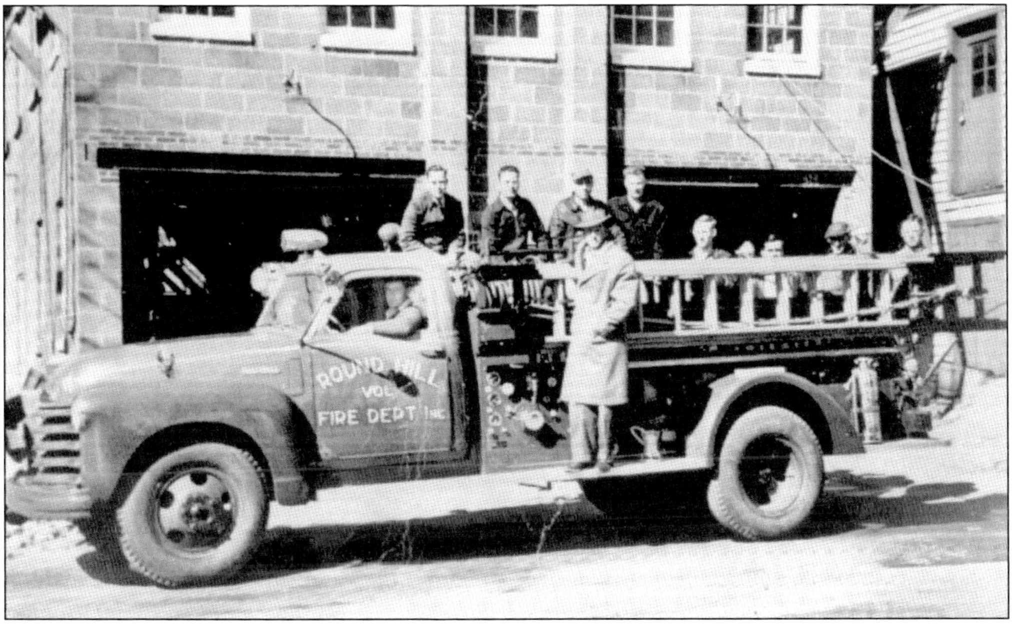

In celebration of the first new fire truck purchased by Round Hill, this 1946 Chevrolet was photographed in front of the department's station on Main Street. Liberty Stable occupied the site before the fire station was built. Tragedy struck the Round Hill community when this fire truck overturned on Route 719 while responding to an emergency. Volunteer firefighter Bob McDaniel was killed. (Courtesy of Round Hill VFD.)

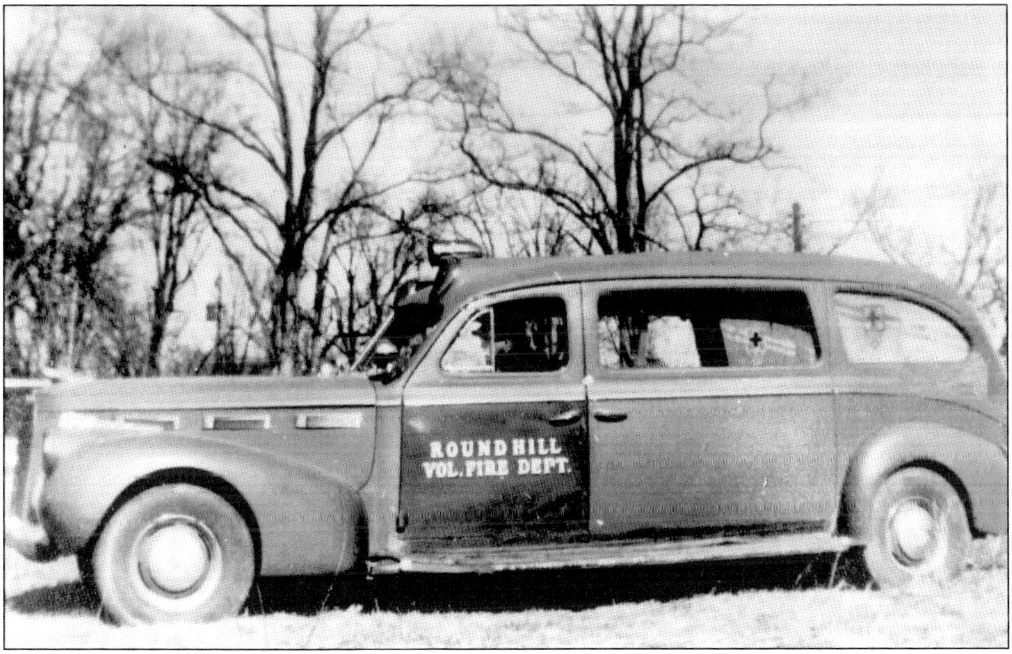

This Round Hill ambulance was one of the first to be operated in Loudoun County. In 1949, this Cadillac ambulance made its way to Round Hill through the owner of the Gulf gas station located in Round Hill. The Cadillac was last seen at a junkyard outside of Philomont, Virginia. (Courtesy of Round Hill VFD.)

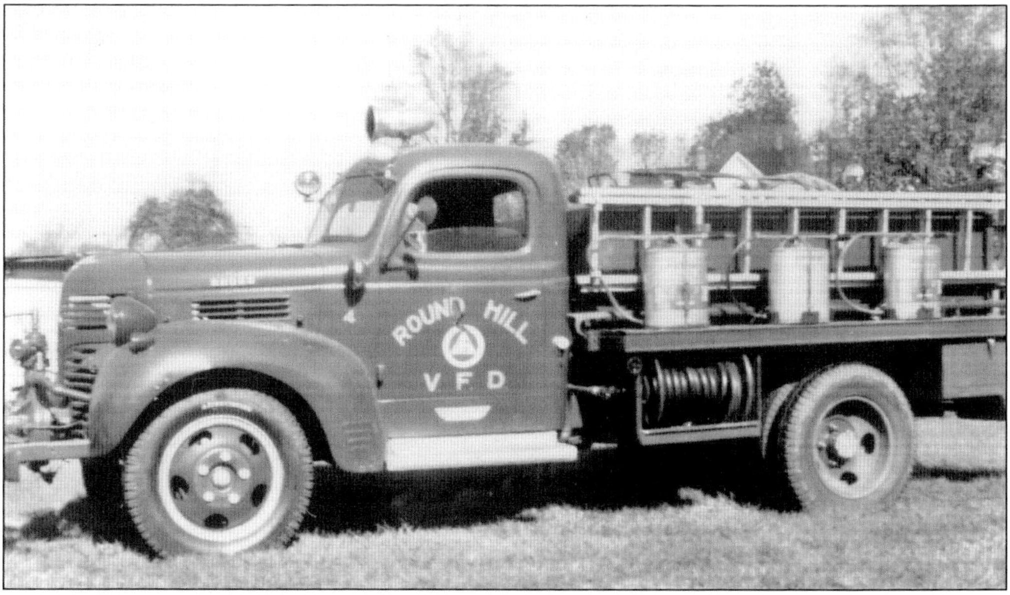

When Round Hill's 1946 Chevrolet fire truck was wrecked, it was replaced with this former Civil Defense unit. The Sterling Park VFD eventually used this 1942 Dodge fire truck when their new fire company became operational. After its time in Sterling, the Dodge was given to the new Lovettsville Fire Department, which had the truck rechassied with a different Dodge cab. (Courtesy of Round Hill VFD.)

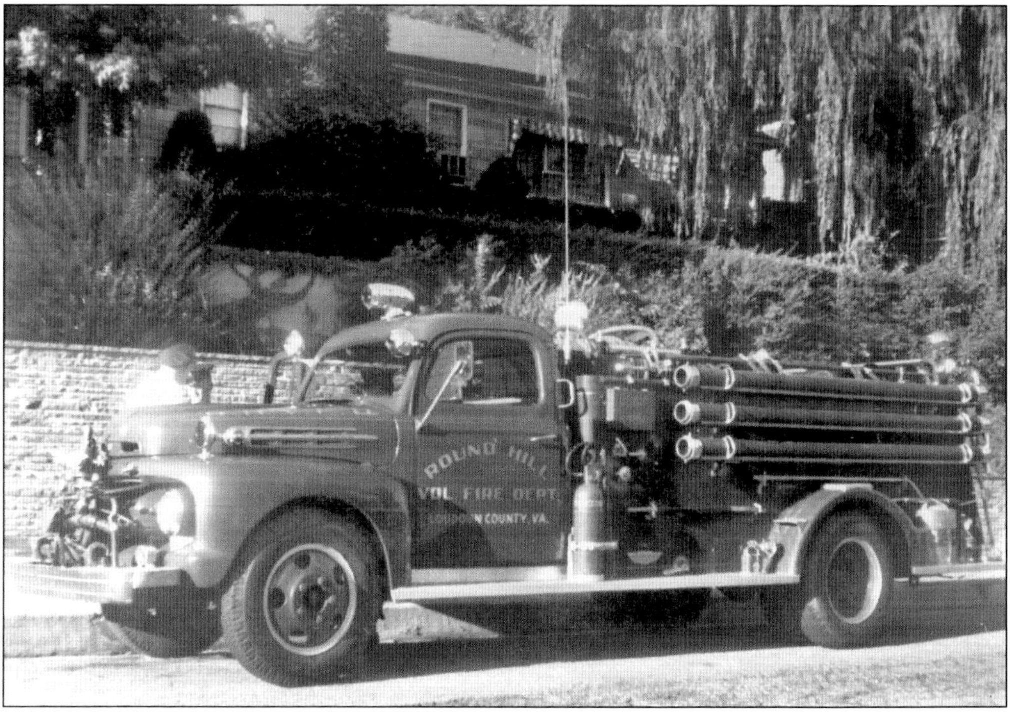

The American Fire Apparatus Company of Battle Creek, Michigan, built this 1952 Ford fire truck for Round Hill. It was equipped with a front-mount pump, a booster reel on top, and three hard suction hoses used for drafting water from ponds or creeks. (Courtesy of Robert Kimball.)

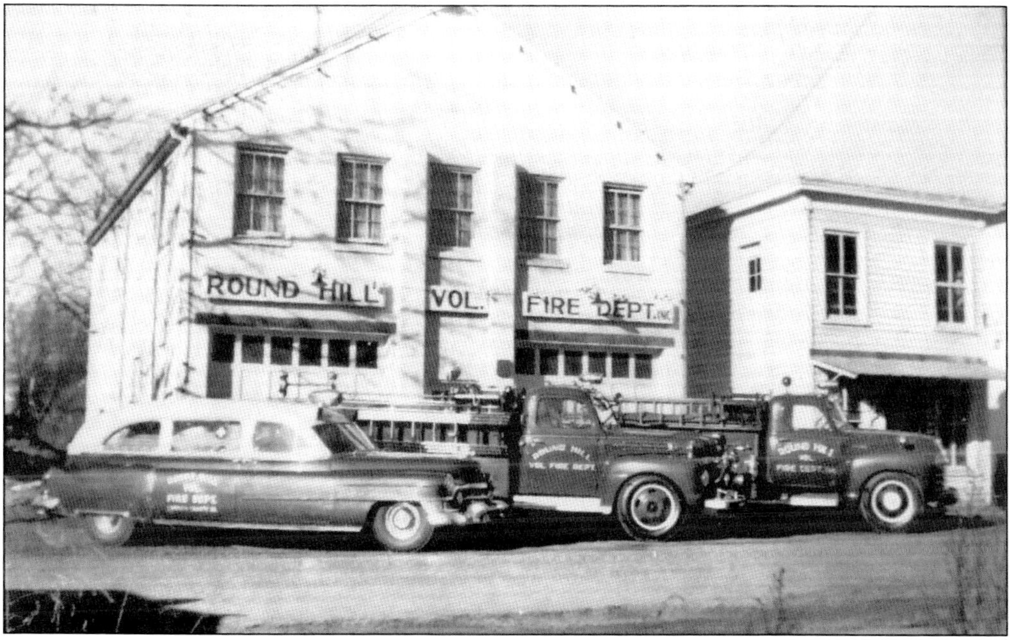

This photograph of the Round Hill Fire Department was taken in the 1950s. To the left is a Cadillac ambulance that Round Hill received with the help of Loudoun Hospital. The 1952 Ford front-mount pumper sits in the middle, and the department's 1946 Chevrolet is on the far right. Station No. 4 has remained at the same location since the department was established. (Courtesy of Round Hill VFD.)

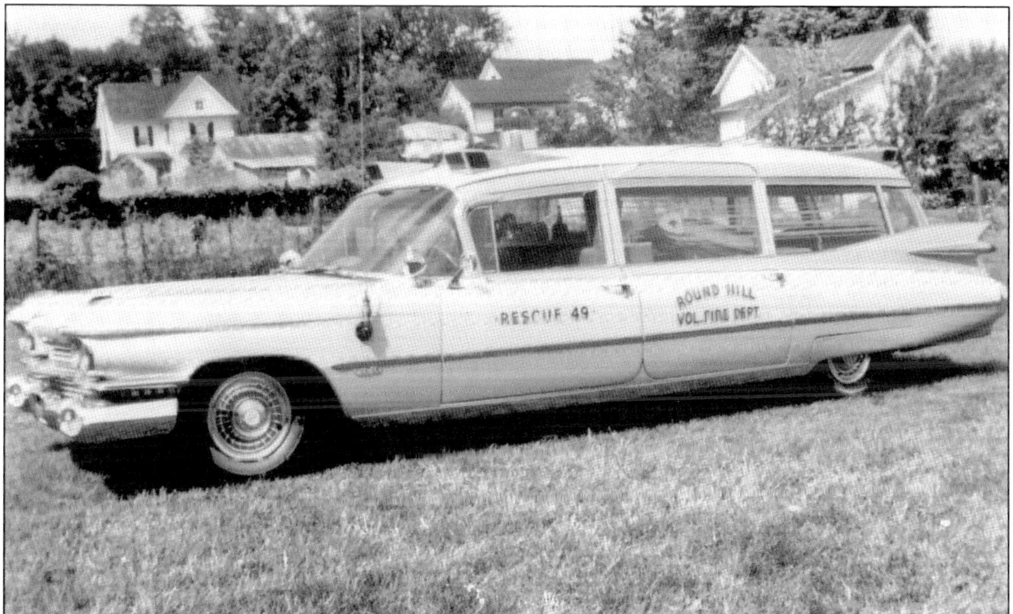

In 1964, Round Hill purchased this used Cadillac ambulance from J. B. Norfleet's Hearse and Ambulance Exchange in Richmond, Virginia. Round Hill originally painted "Rescue 49" red and white but later repainted the ambulance white and green. Miller Meteor, an ambulance manufacturer, built the ambulance body on the Cadillac. (Courtesy of Round Hill VFD.)

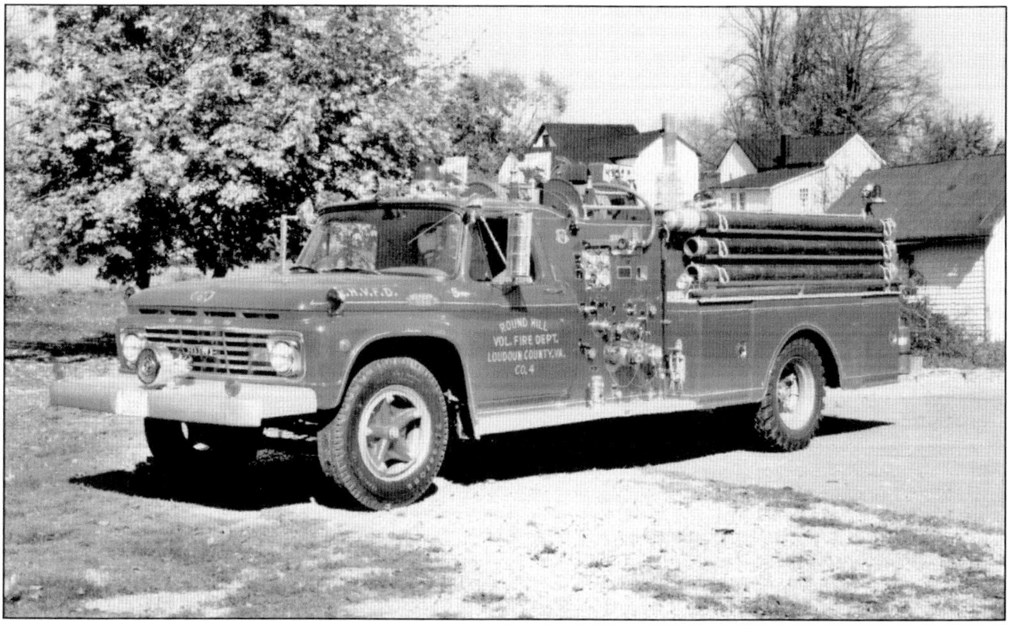

This red Ford fire truck was built by Howe Fire Apparatus Company and purchased new by Round Hill in 1963. Engine No. 4 was equipped with a 750-gallon-per-minute pump and held 500 gallons of water. Note the extended front bumper with the Federal Q siren. Round Hill later sold this fire truck to a citizen in Neersville, who used the truck on his farm. (Courtesy of Chuck Madderom.)

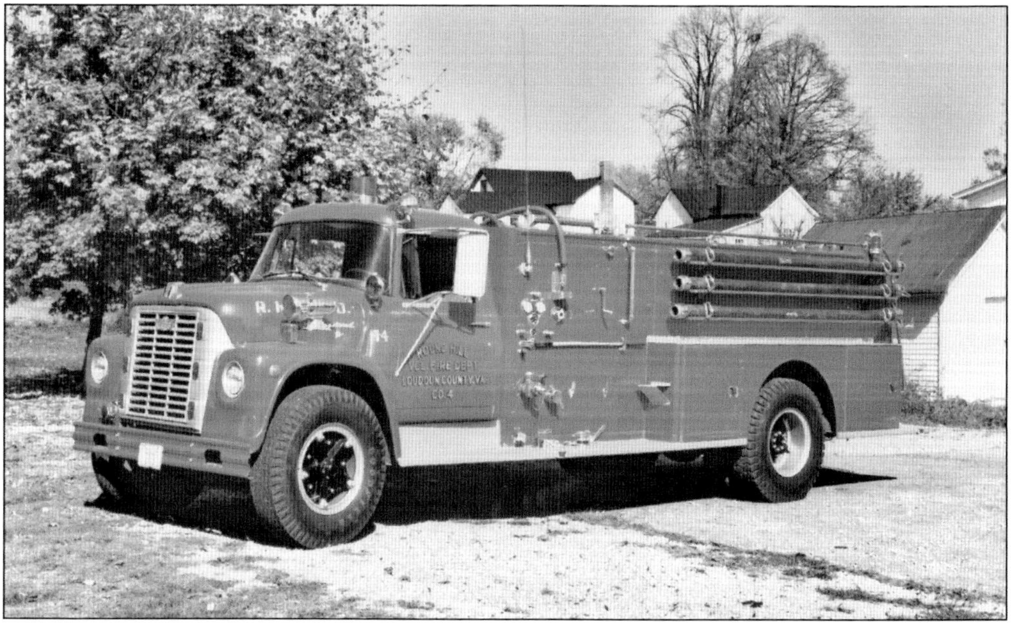

Round Hill went back to the American Fire Apparatus Company to have a new tanker constructed. The members purchased a new 1969 International Loadstar cab and chassis from Loudoun Truck Center and had it sent to American for the building of a fire-truck body. The result was the new Tanker No. 4, a red fire truck capable of carrying 1,500 gallons of water. (Courtesy of Chuck Madderom.)

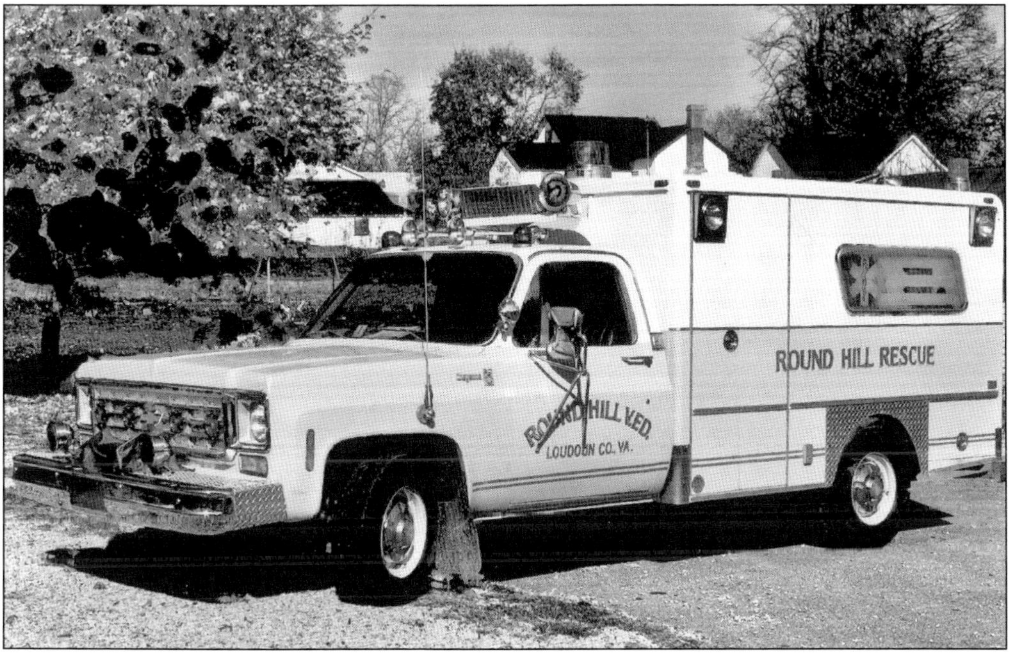

Round Hill received this 1979 Chevrolet ambulance, which was built by the Swab Wagon Company in Pennsylvania. During the 1970s, a study was published that concluded that lime-green apparatus were more visible to the public and thus safer when responding to an emergency call. Round Hill's new delivery arrived painted lime green and white. (Courtesy of Robert Kimball.)

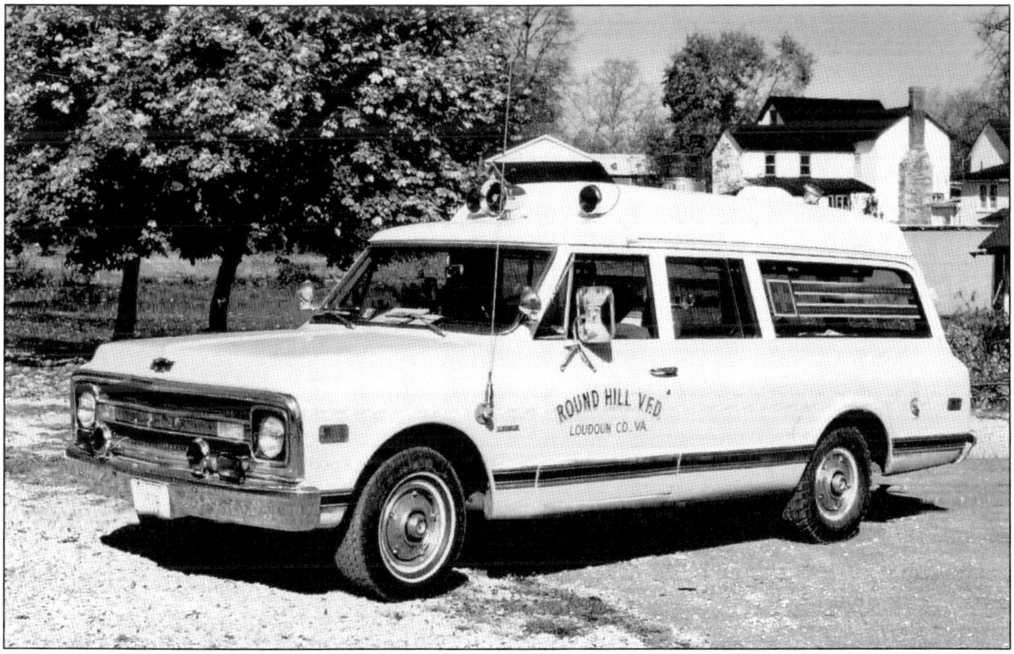

In 1969, the Round Hill VFD received a Chevrolet Carryall, which they used as an ambulance. It was originally painted red and white. When the department received its new 1979 Chevrolet/Swab ambulance in the new lime-green color, they decided to repaint the 1969 Chevrolet to match the other ambulance. (Courtesy of Robert Kimball.)

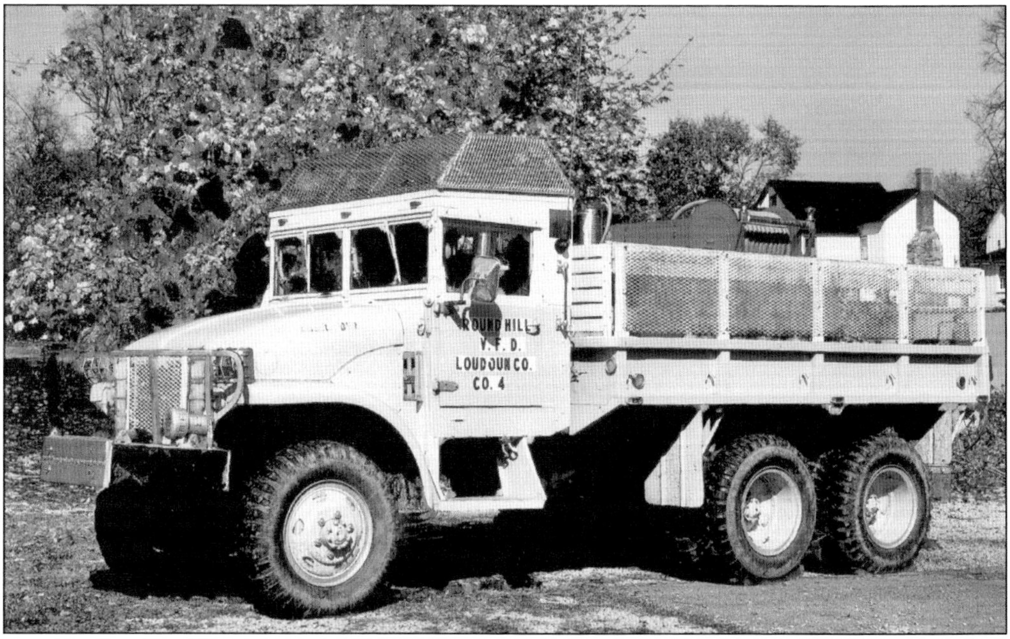

This 1950s surplus military truck was purchased by Round Hill during the 1970s and repainted in the department's "new" colors of lime green and white. Members removed the canvas top and added a roof and screen protection for the warning lights on the roof. This truck replaced an earlier single-axle military truck that was painted yellow and that burned during a mountain fire behind the Hillsboro Elementary School. (Courtesy of Robert Kimball.)

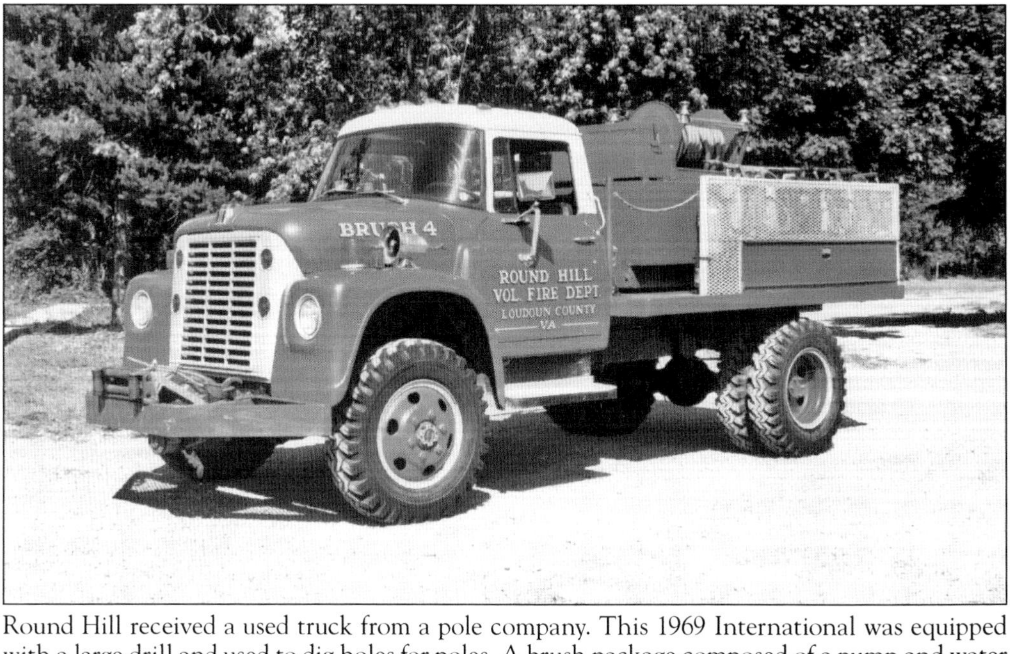

Round Hill received a used truck from a pole company. This 1969 International was equipped with a large drill and used to dig holes for poles. A brush package composed of a pump and water tank was purchased from Slagle's in Southern Virginia. Chase Schneider and Fenton Simpson from Round Hill rebuilt the truck into a very durable brush unit. Note the lack of warning lights on the cab roof.

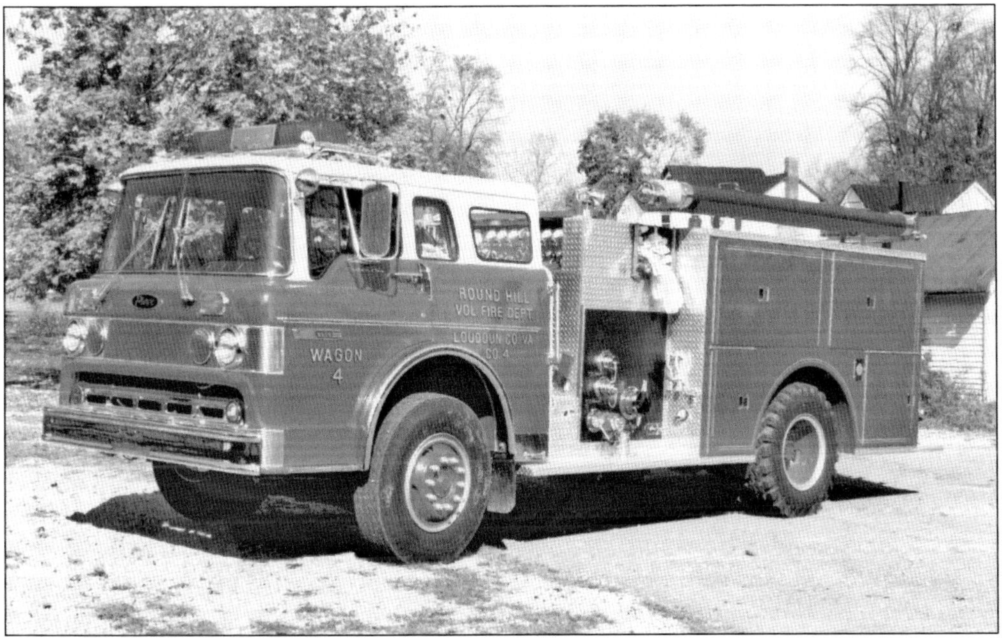

Round Hill members purchased the first top-mount pump panel fire truck in Loudoun County. The driver operates the pump on the truck behind the cab rather than on the ground. In 1979, Wagon No. 4 received this Ford C 1,000-gallon-per-minute pumper built by Pierce. It was painted red and white, as the department no longer ordered apparatus painted lime green and white. (Courtesy of Chuck Madderom.)

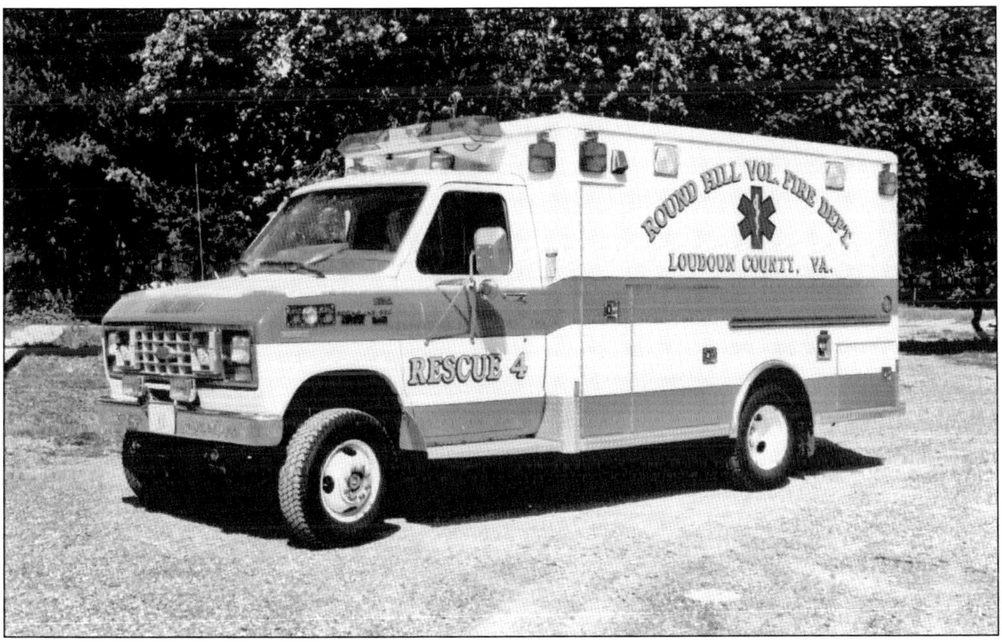

A variety of ambulances built by Marque or Wheeled Coach on Ford E350s were used by Round Hill. These ambulances were painted white with two large red stripes. Rescue No. 4-1 was a 1988 Ford Econoline 350 built by Wheeled Coach. It was later sold to the Lucketts VFD in 2003.

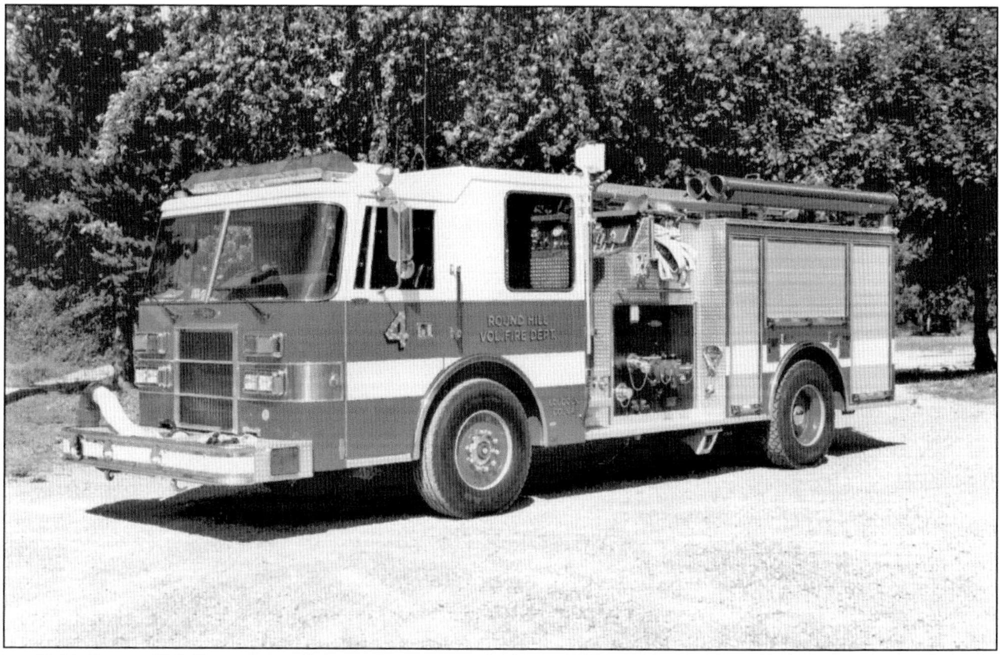

In 1996, Round Hill purchased a 1994 Pierce Dash fire truck. Prior to its arrival at Round Hill, Pierce used this truck as a demonstrator model. Wagon No. 4 came equipped with a top-mounted, 1,250-gallon-per-minute pump and had an unusual four-door cab. Unlike most four-door pumpers with the doors on the side, the rear doors of Wagon No. 4 are behind the cab near the pump panel.

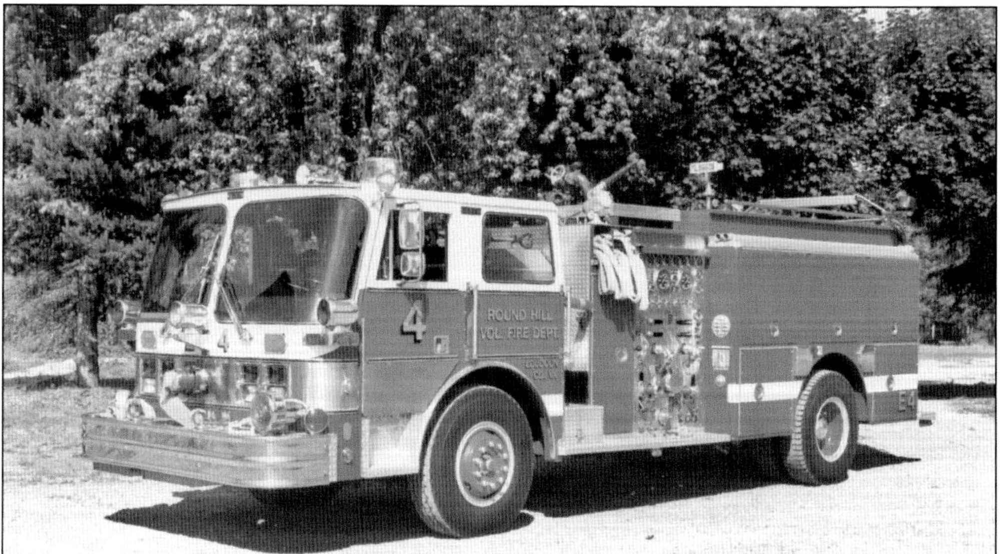

Round Hill was in need of a back-up pumper in 1999, so the membership voted to purchase a used 1985 Hahn from the Stonewall Jackson VFD, located near Manassas, Virginia. The Hahn was taken to LSI Fire, a fire truck dealer and builder located in Purcellville, Virginia, which readied the pumper for service and repainted the Hahn from white and yellow to red and white. In 2006, the Hahn was given to the Loudoun County Fire Rescue Training Center.

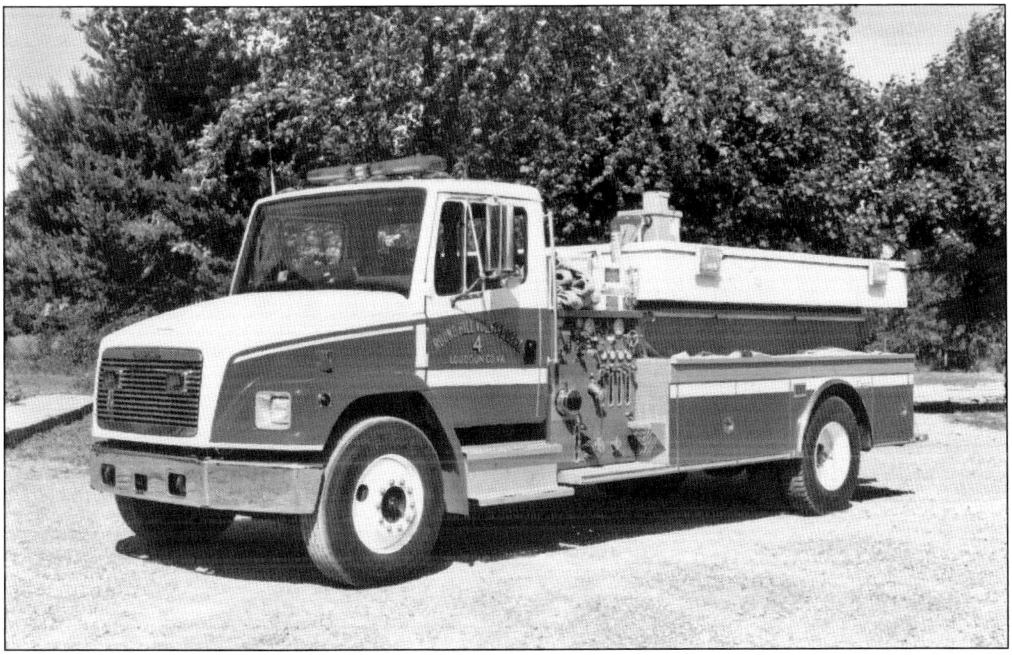

With the 1969 International/American tanker showing its age, the Round Hill VFD purchased a new tanker truck built by Superior on a 1996 Freightliner 70 cab. Tanker No. 4 had a lower tank body and held 1,800 gallons of water. The tanker, which is painted red and white, was replaced in 2007.

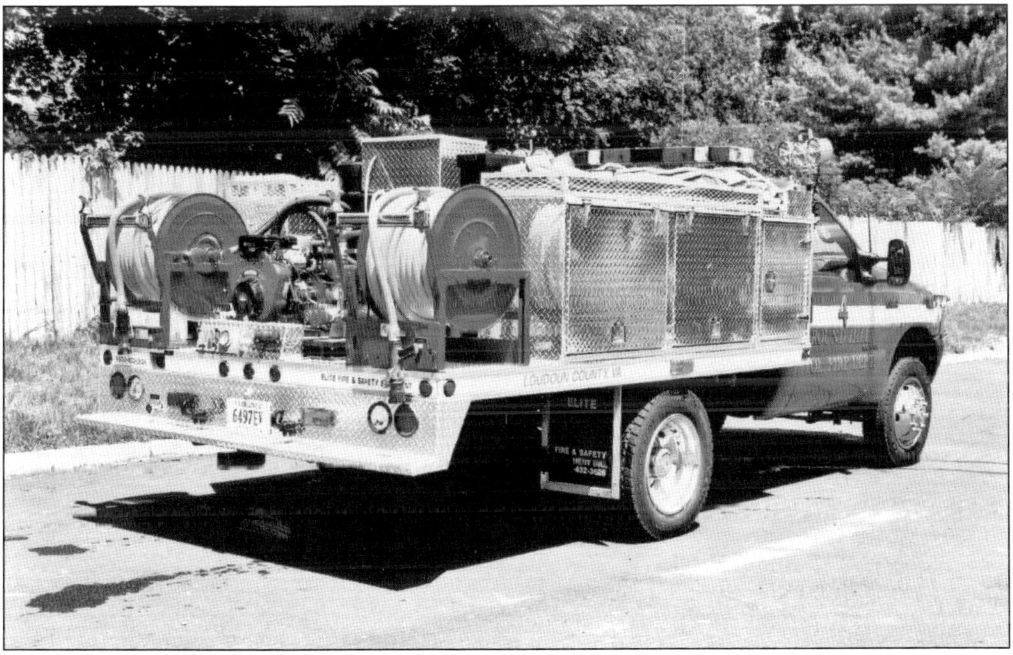

Round Hill's 1969 International brush truck was replaced in 2002. Brush No. 4 was built by Elite in Florida on a Ford and came equipped with a 150-gallon-per-minute pump and held 300 gallons of water. Brush No. 4 was built upon a flatbed-truck body, a popular option for brush trucks in Loudoun County.

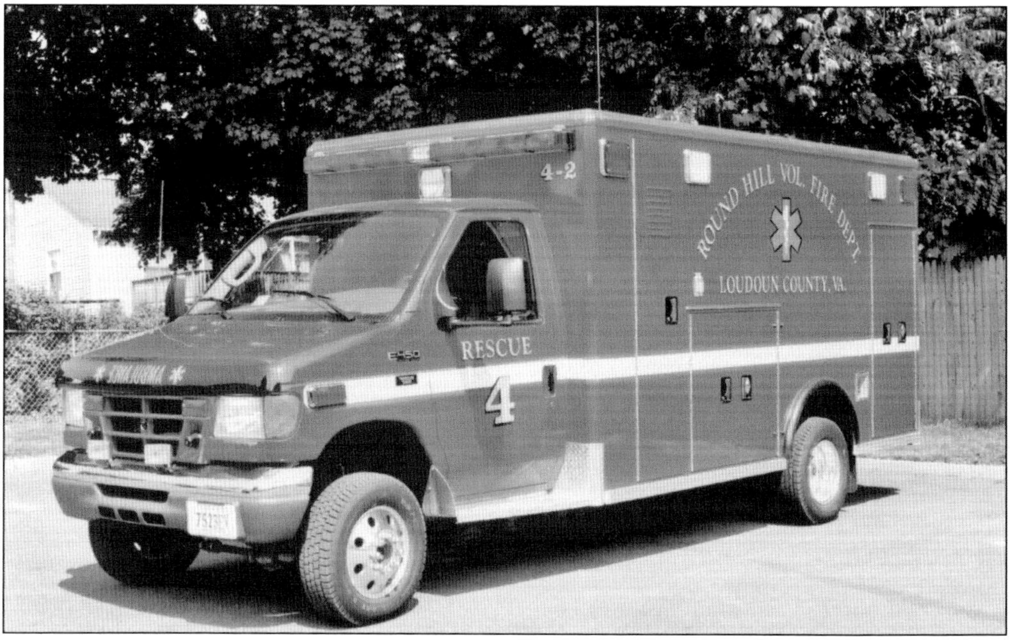

Round Hill's 2004 ambulance came delivered with a new variation in color. Rescue No. 4-2 is a Ford E350 built by Wheeled Coach. The all-wheel-drive ambulance is designed for inclement weather and is painted a deeper color of red than Round Hill's other fire and rescue apparatus.

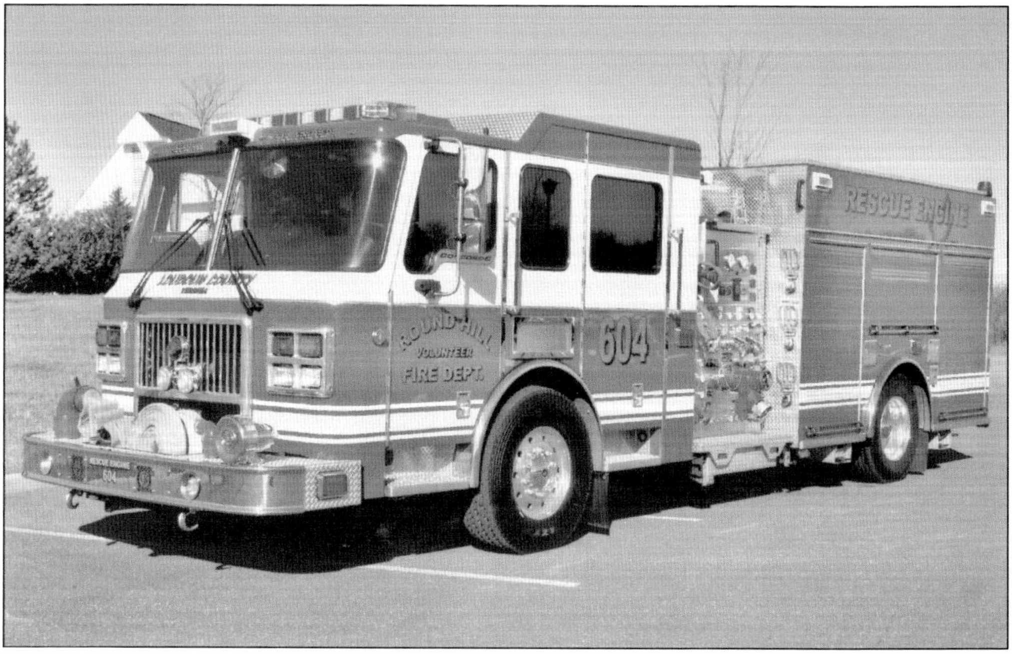

Rescue Engine No. 4 arrived in Round Hill during 2005 and is a Seagrave Concorde. The department purchased a rescue engine so that extrication equipment could be carried. Painted red and white and complete with roll-up doors, Rescue Engine No. 4 is equipped with a 1,500-gallon-per-minute pump and carries 750 gallons of water and 50 gallons of foam.

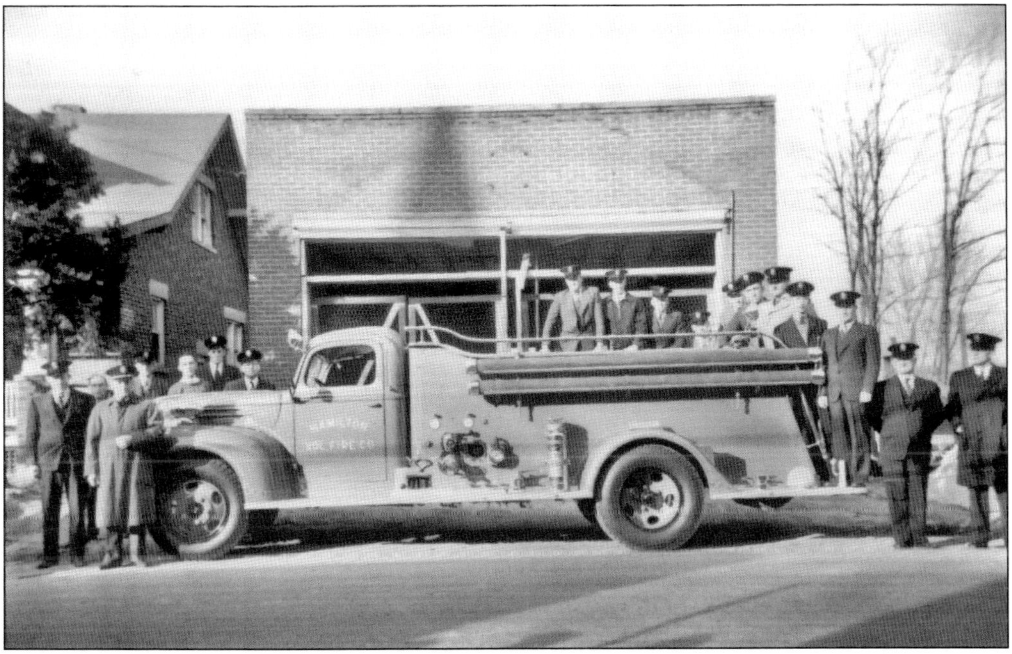

Hamilton Fire Station No. 5 was established in 1944, and the fire station was an old dime store. As shown in this photograph, chicken wire attached to a wood frame was used as the garage doors. This is Hamilton's first fire truck, a 1945 Ford built by Oren. The truck, plus firefighting equipment, cost $5,362 when ordered. (Courtesy of Hamilton VFD).

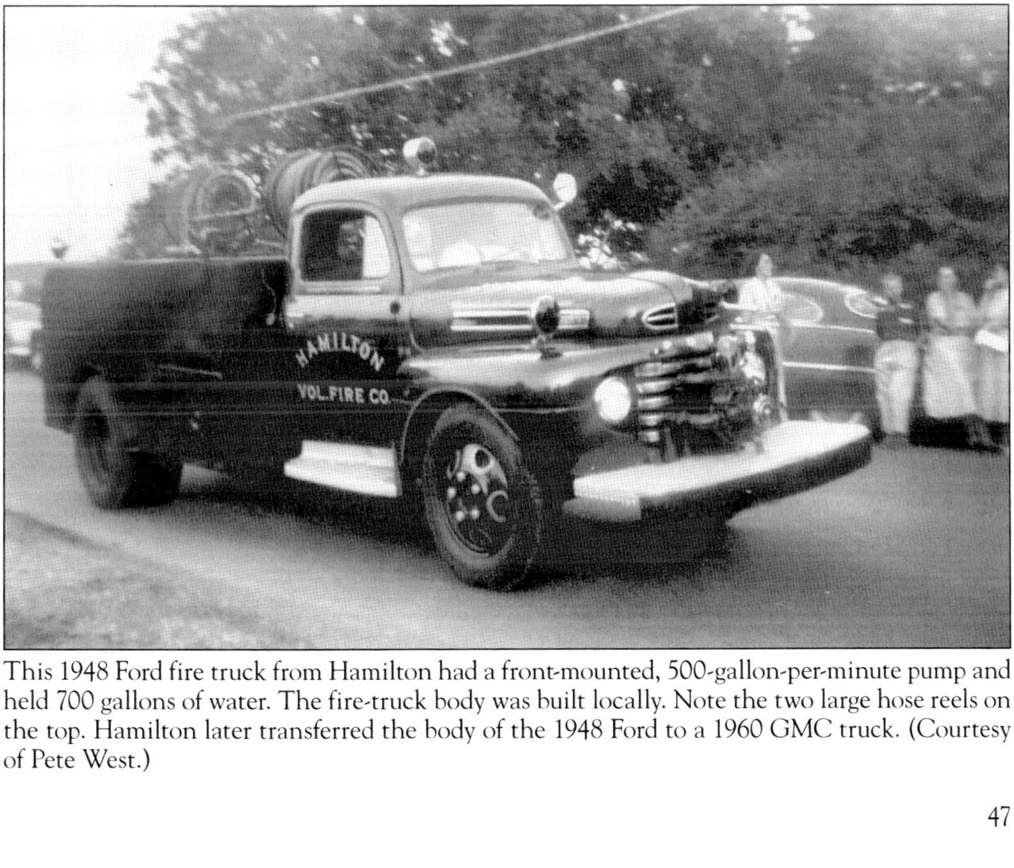

This 1948 Ford fire truck from Hamilton had a front-mounted, 500-gallon-per-minute pump and held 700 gallons of water. The fire-truck body was built locally. Note the two large hose reels on the top. Hamilton later transferred the body of the 1948 Ford to a 1960 GMC truck. (Courtesy of Pete West.)

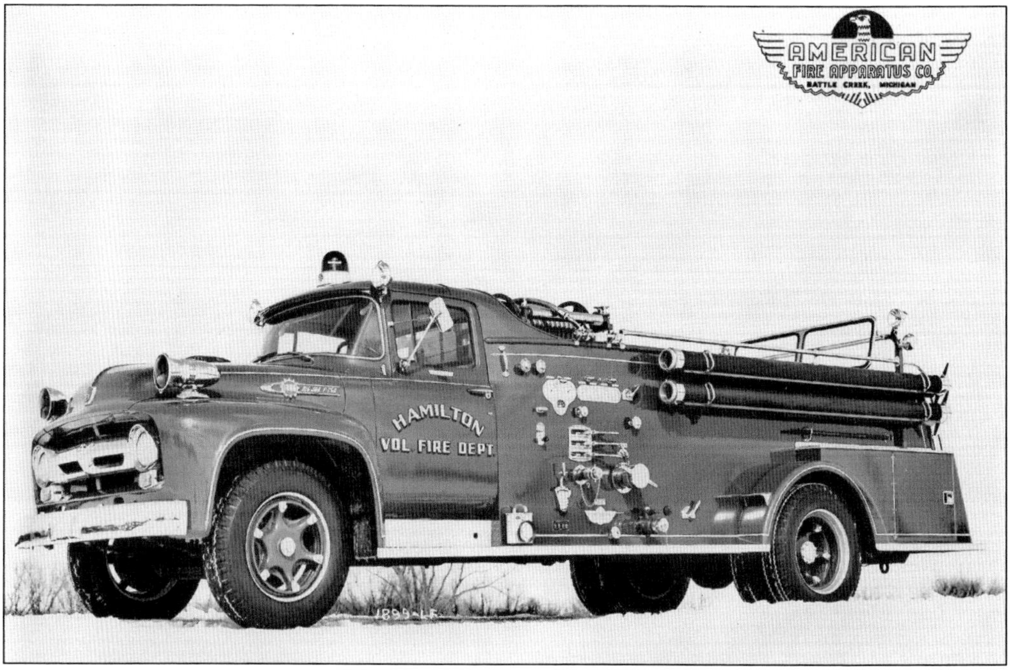

The Glenn D. Culbert Company sold Hamilton their 1956 Ford Big Job F-750, built by the American Fire Apparatus Company. This is the American delivery photograph. This Ford would later see service in Round Hill and Neersville. Today the truck is stored in a barn in Hamilton, hopefully awaiting a restoration.

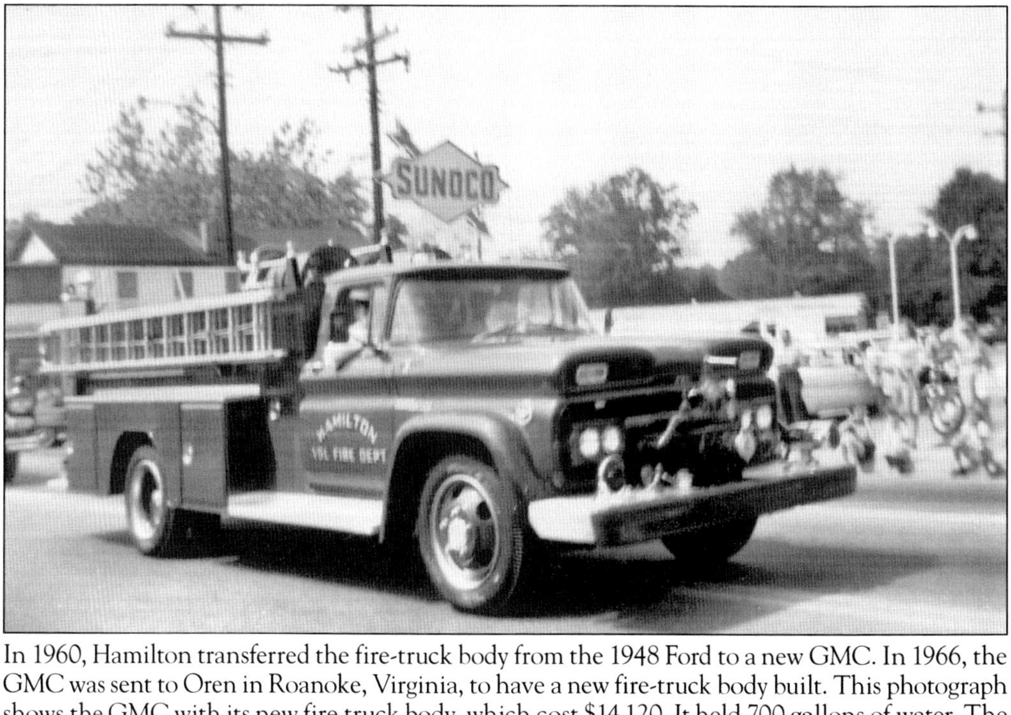

In 1960, Hamilton transferred the fire-truck body from the 1948 Ford to a new GMC. In 1966, the GMC was sent to Oren in Roanoke, Virginia, to have a new fire-truck body built. This photograph shows the GMC with its new fire truck body, which cost $14,120. It held 700 gallons of water. The GMC was later sold to Station No. 12 in Lovettsville. (Courtesy of Pete West.)

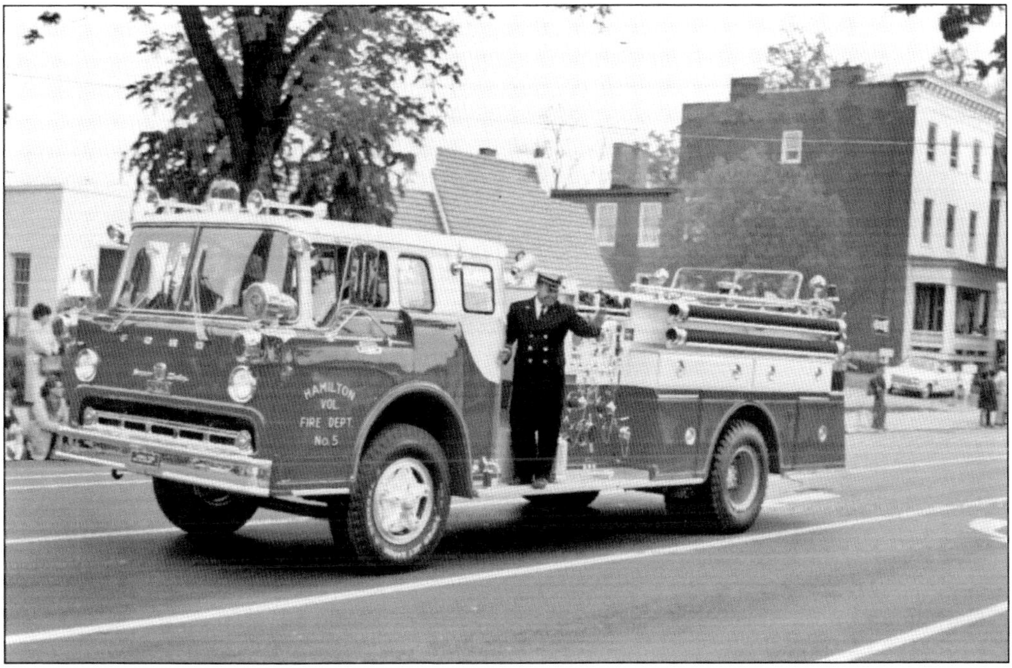

Oren built Hamilton's 1967 Ford C fire truck, and it established a trend with departments in Loudoun County. First, the Ford C had a cab that allowed firefighters to be seated behind the driver and the officer, rather than riding on the back step. In addition, Oren built the fire truck with cross-lays, which are the hose loads just above the pump panel. Future apparatus in Loudoun followed Hamilton's design. It was later sold to Tazewell, Virginia.

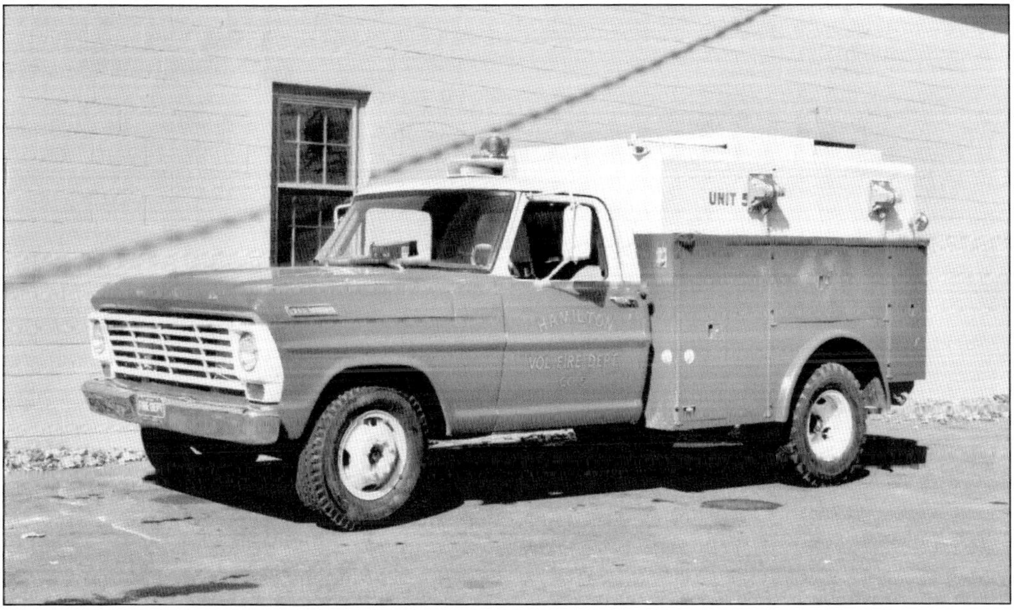

Unit No. 5 was donated to the Hamilton VFD by C&P Telephone Company and was a 1967 Ford with a utility body built by Baker. It was originally an equipment truck for Hamilton, and the department later added an air cascade system. The 1967 Ford would later be replaced with another used telephone truck. (Courtesy of Chuck Madderom.)

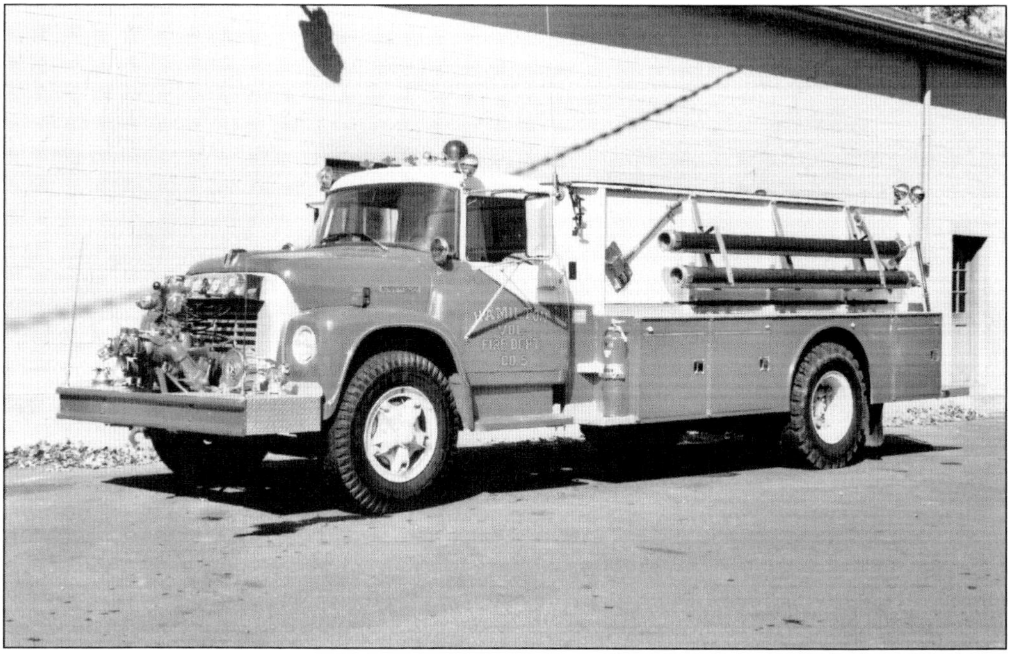

Tanker No. 5 from Hamilton was originally a 1973 International with a 1,200-gallon tanker built by Atlas. Hamilton had numerous problems with the Atlas body, and as a result, the body was replaced in 1979 with a new 4-Guys 1,500-gallon tanker body. Tanker No. 5 was sold to the Rapidan Volunteer Fire Department in Culpepper County, Virginia. (Courtesy of Chuck Madderom.)

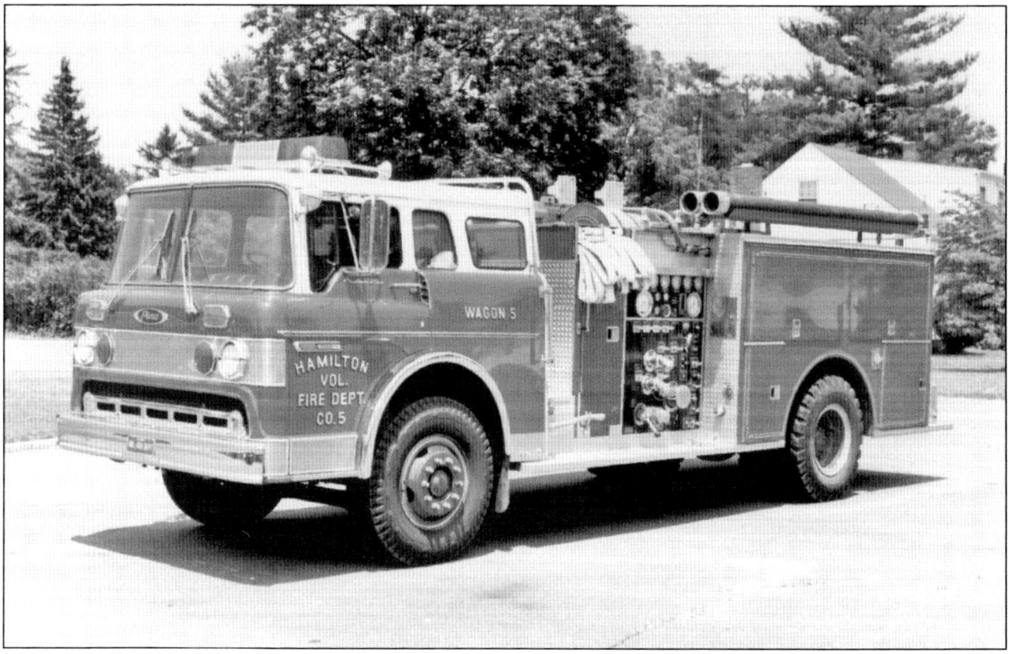

In 1978, Hamilton received a Pierce fire truck built on the popular Ford C cab and chassis. During the late 1970s and early 1980s, Loudoun County departments ordered a variety of these Ford C engines. Wagon No. 5 was equipped with a 1,000-gallon-per-minute pump and carried 750 gallons of water. After service in Hamilton, it was sold to a fire department in Alabama.

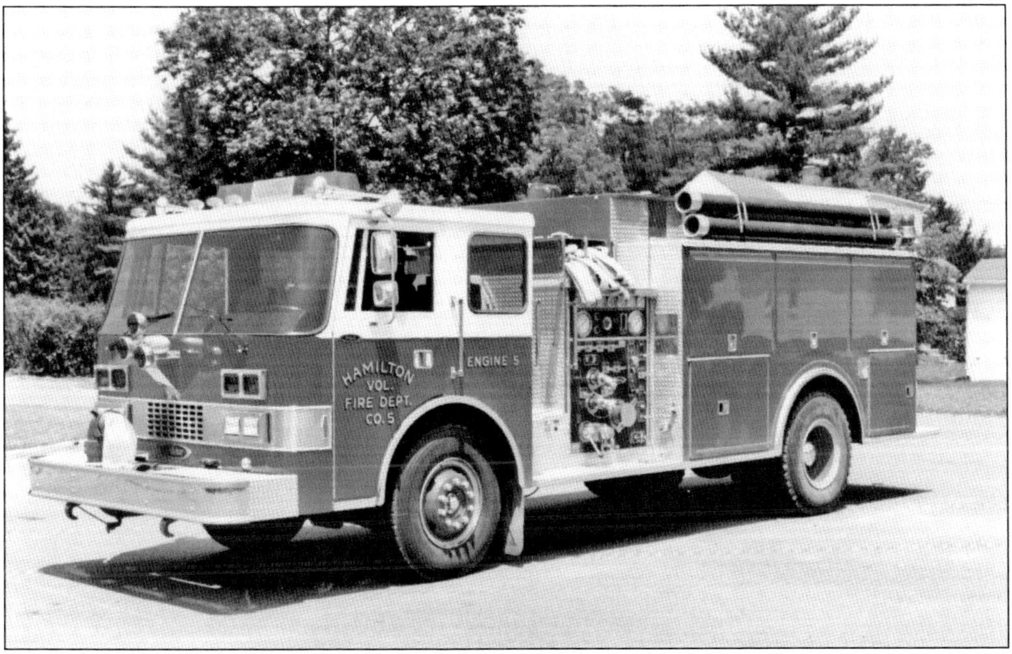

When Hamilton was ready to replace the 1973 International running as Tanker No. 5, they found that a Pierce custom tilt cab fire truck was in the same price range as a commercial chassis truck. The department purchased this 1984 Pierce Dash with a 1,500-gallon water tank. Compared with a commercial cab truck, the Dash cab had the capacity to seat more firefighters.

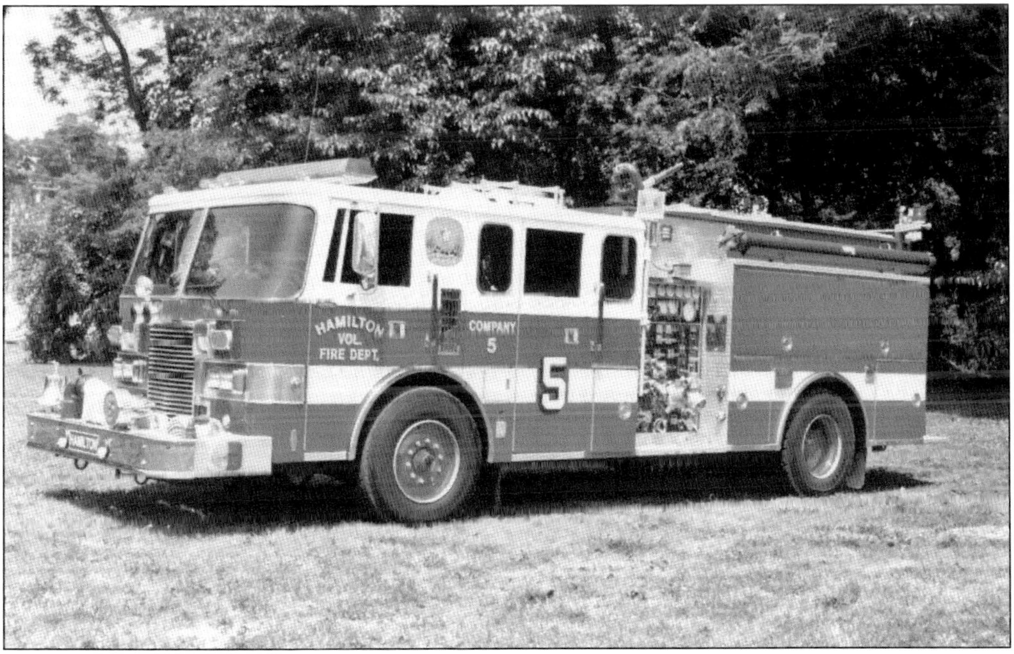

Since 1978, the Hamilton volunteers have purchased fire trucks from Pierce Manufacturing. When this 1989 Pierce Lance was delivered to Hamilton, it ran as Wagon No. 5 and was the department's first four-door cab. Wagon No. 5 is equipped with a 1,250-gallon-per-minute pump and carries 750 gallons of water.

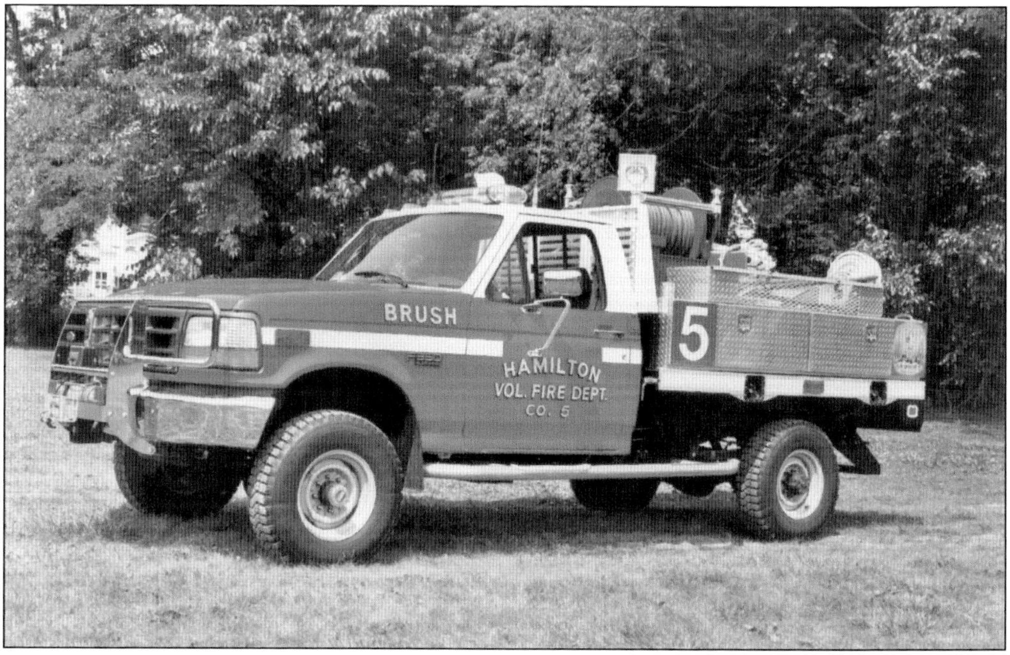

Brush No. 5 for Hamilton is a unique 1994 Ford F-350 built locally by LSI Fire of Purcellville. Designed on a flatbed body, the brush truck carries 250 gallons of water, as well as a variety of compartments to store firefighting equipment. It would later be upgraded with larger tires and a new flatbed body.

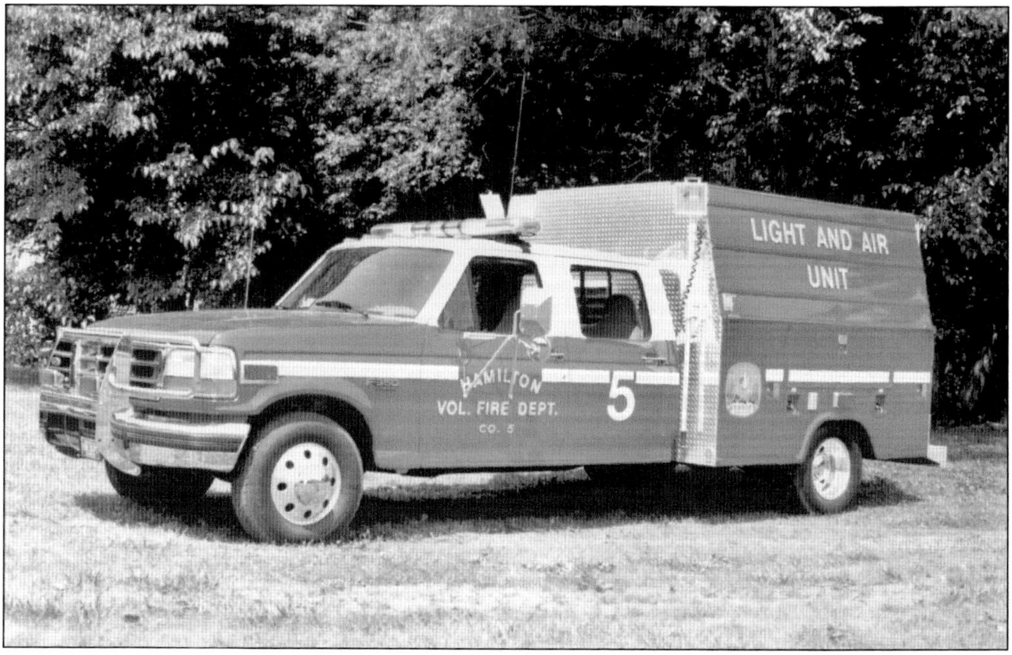

For many years, Hamilton's Light and Air No. 5 has handled the duties of refilling air bottles at the scene of large incidents, as well as providing additional lights when needed. This unusual 1995 four-door Ford F-350 has a Stahl utility box, and it replaced a 1980 Chevrolet equipment truck that was donated to Hamilton by C&P Telephone Company.

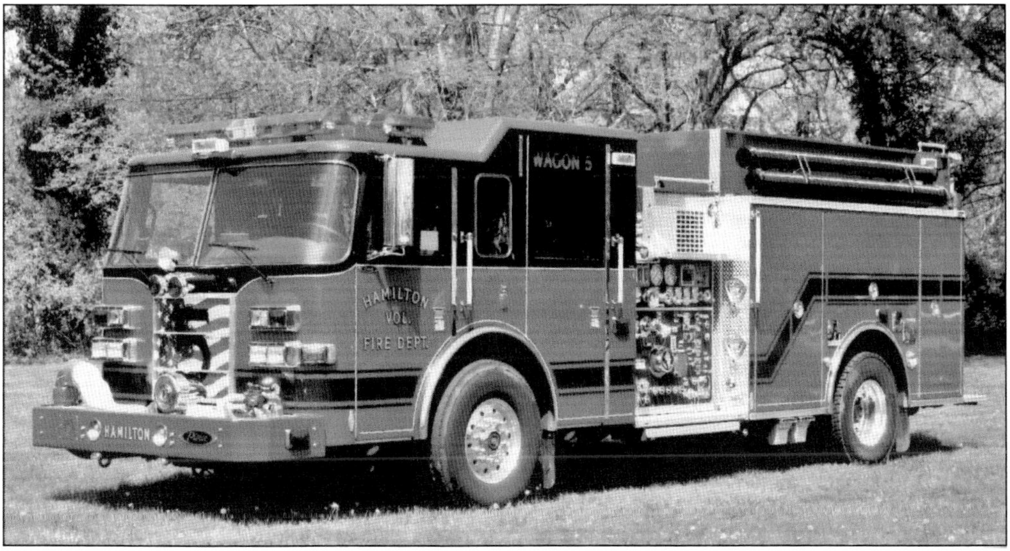

An impressive new Wagon No. 5 was delivered to Hamilton in 2006. Built by Pierce with a Dash cab, Wagon No. 5 carries 1,000 gallons of water and is equipped with a 1,500-gallon-per-minute pump. The Pierce arrived at Hamilton painted red and black, the new colors of the department. A unique feature of the Pierce is that the front bumper is painted red.

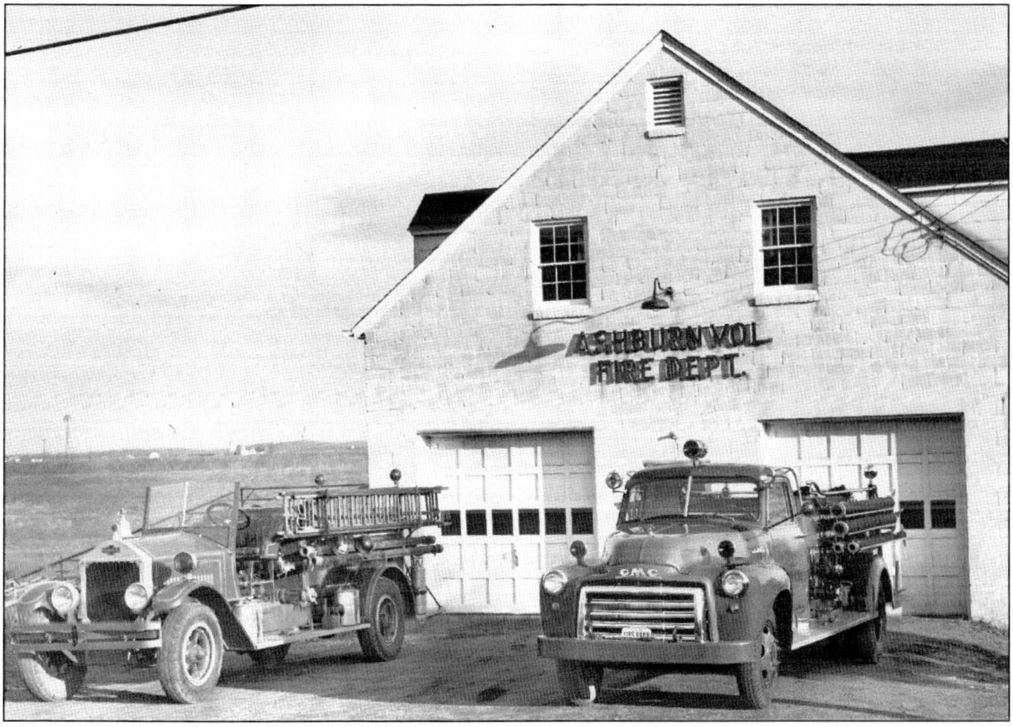

Ashburn's first fire truck was this used 1928 American LaFrance purchased for $3,225 (left). It was sold in 1955 to a local farmer and was used to water cattle. The department kept the bell from this truck, and it is still used by the department for special ceremonies. Ashburn's 1948 GMC/American (right) sits next to the American LaFrance in front of Station No. 6. (Courtesy of Ashburn Volunteer Fire and Rescue Department)

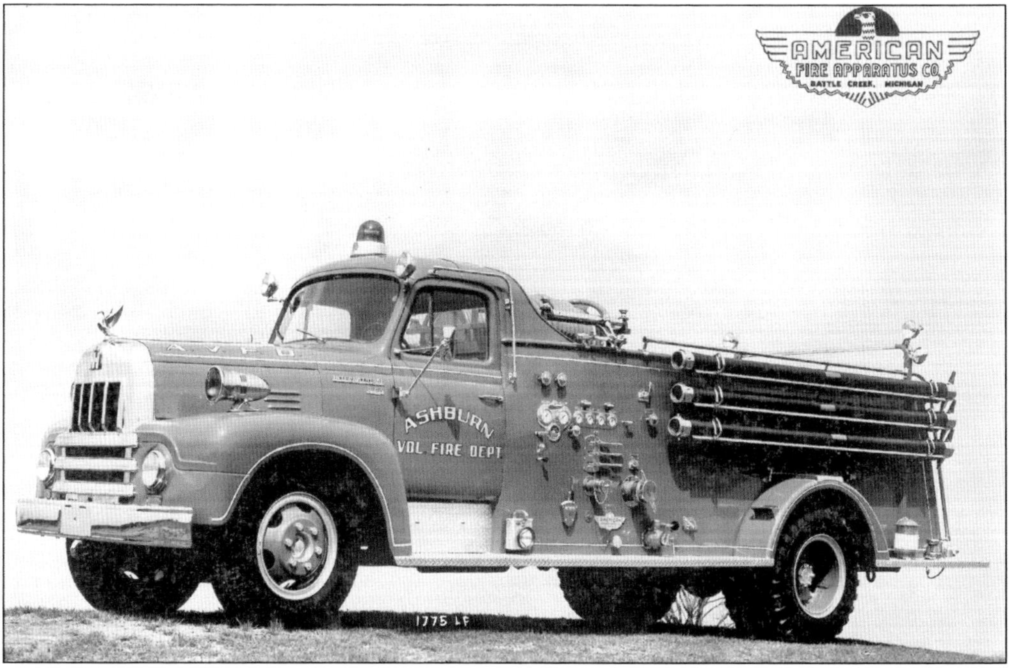

This delivery photograph shows Ashburn's 1955 International, which was built by American with a 500-gallon-per-minute pump and a 500-gallon water tank. It carried 1,000 feet of 2½-inch supply hose. With its chrome grill, it won numerous trophies at fire department parades. After serving more than 20 years at Ashburn, the International was sold in 1977 to a fire-truck dealer located in Alabama.

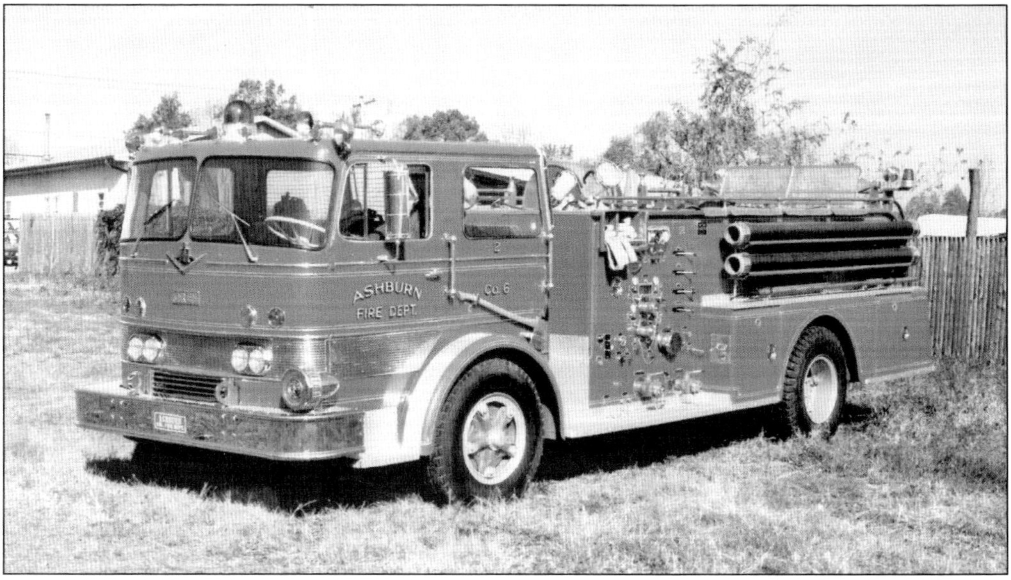

In 1967, Ashburn placed into service a red 1967 International equipped with a 1,000-gallon-per-minute pump. Oren, a popular fire-truck manufacturer located in Roanoke, Virginia, built this fire truck for the Ashburn community. This was Ashburn's last red fire truck. In 1984, this International was sold to the Neersville VFD, also located in Loudoun County. (Courtesy of Chuck Madderom.)

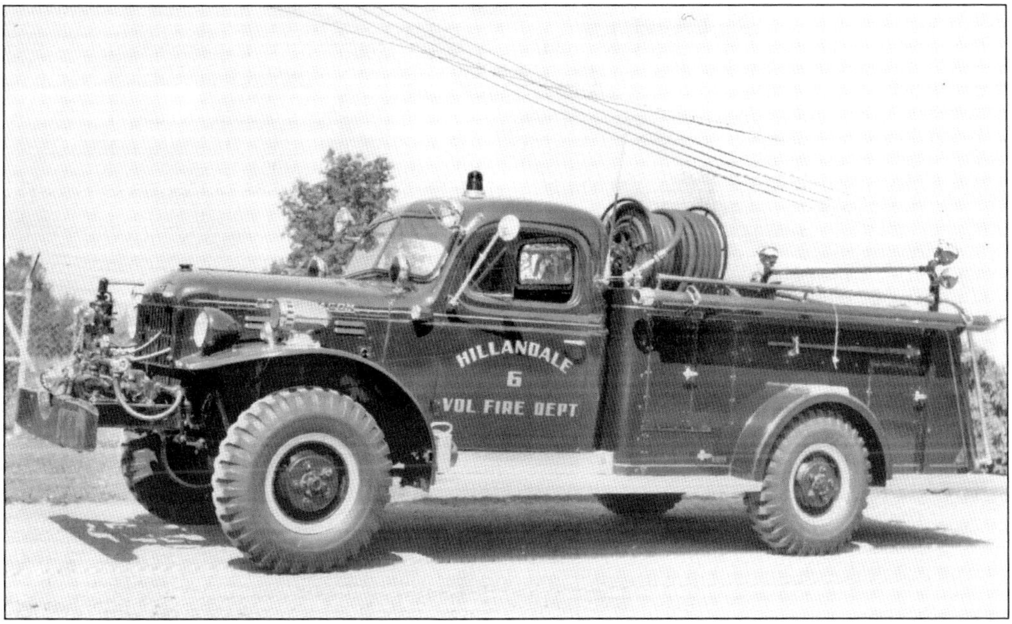

Ashburn once owned this 1952 Dodge Power Wagon brush truck. Ashburn was the third department to own the Dodge. The Hillandale, Maryland, VFD in Montgomery County (shown above) originally ordered the brush truck. Hillandale sold the Dodge to the Jefferson VFD in Fairfax County, Virginia. From Jefferson, it went to Ashburn. Painted red, it was equipped with a 500-gallon-per-minute pump and a large booster reel.

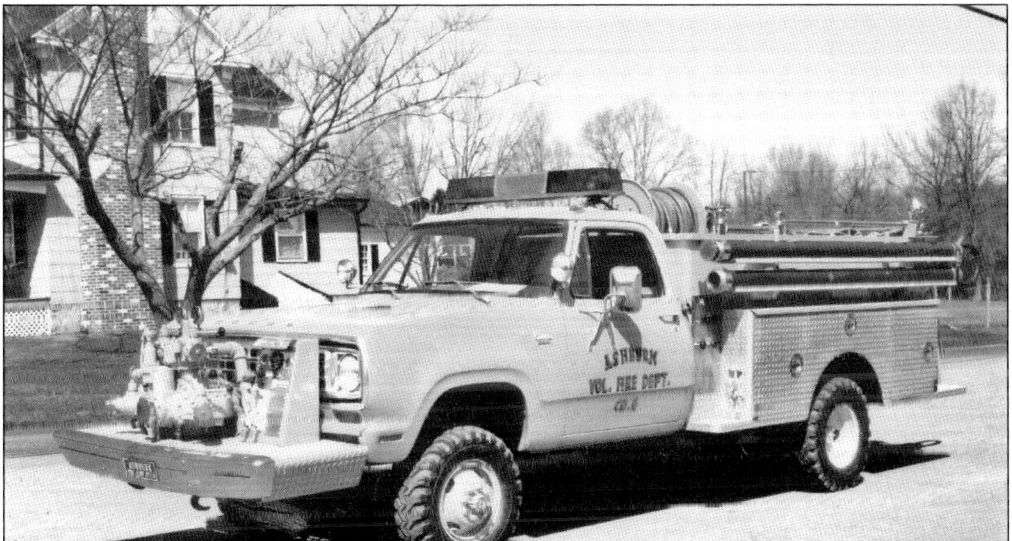

The first yellow fire truck purchased by Ashburn was a mini-pumper built by Stinebaugh and Sons, a small fire-truck manufacturer in West Virginia. This 1974 Dodge had a front-mount, 500-gallon-per-minute pump and carried 300 gallons of water. This Dodge/Stinebaugh was later sold to the Bergton, Virginia, Volunteer Fire Department in Rockingham County, where it was still in service in 2007.

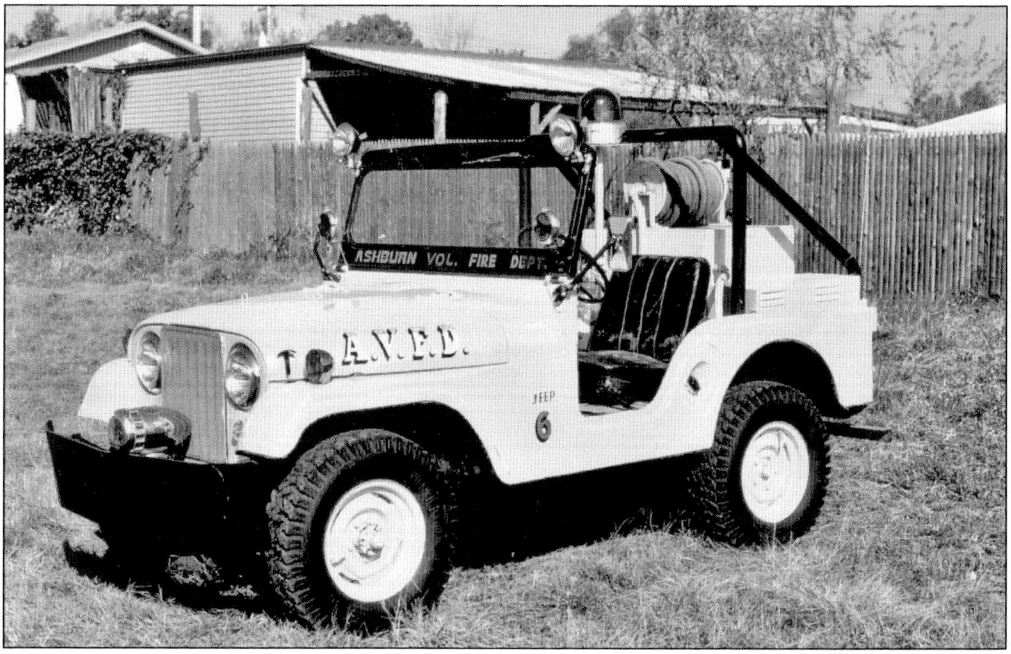

Ashburn obtained this older Jeep from a fire department in Fairfax County, Virginia, and painted it yellow and white. Ashburn's Jeep had the radio designation of "Jeep 6" and was later replaced by a 1977 Jeep that Ashburn purchased from the Sterling Volunteer Fire Department. (Courtesy of Robert Kimball.)

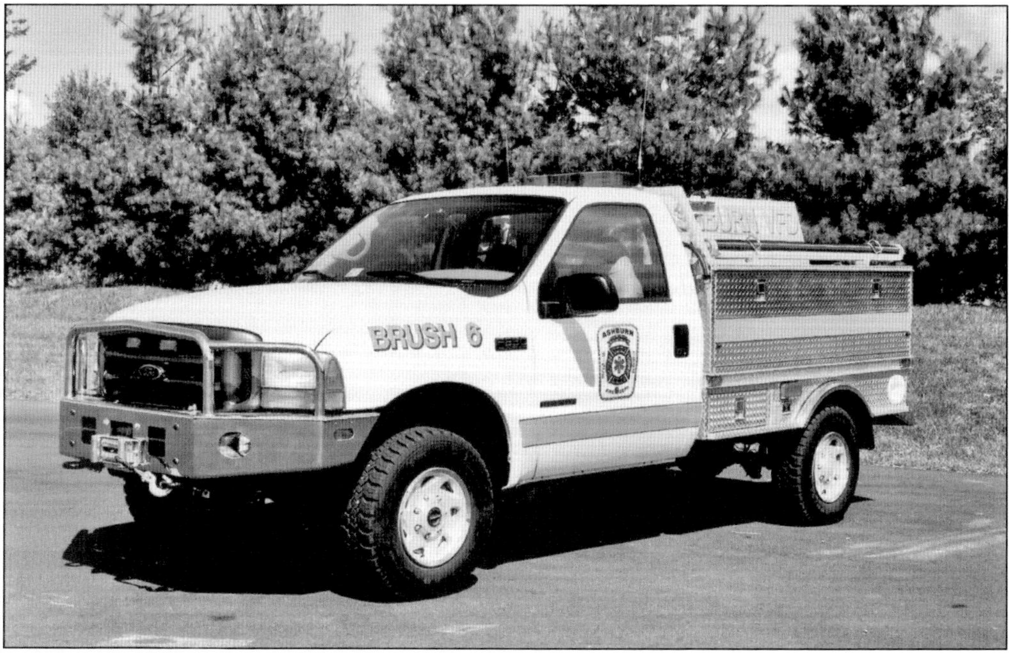

Ashburn's 1977 Jeep brush truck was replaced in 2000 by a new brush truck built by Ashburn's mechanic, Dave Weddle. Brush No. 6, a 2000 Ford F-350, has a rear-mounted, 400-gallon-per-minute pump and holds 250 gallons of water. Brush No. 6 has two large compartments to hold a variety of equipment needed to fight brush and outside fires.

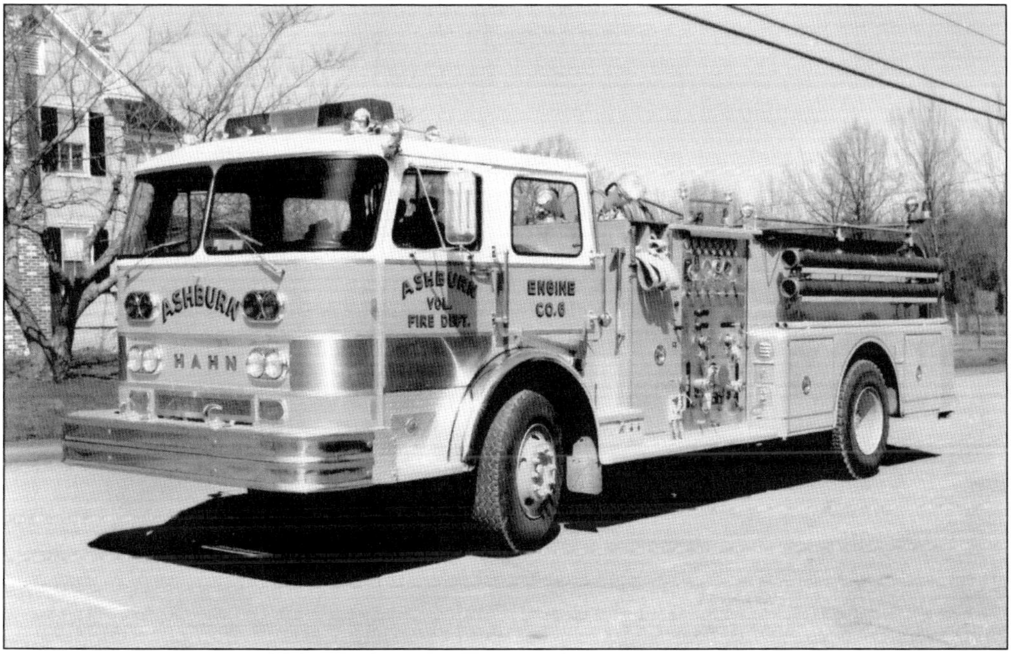

The first custom-built fire truck for Ashburn was delivered in 1980 and was built by Hahn, a fire-truck manufacturer located in Hamburg, Pennsylvania. The Hahn had a 1,000-gallon-per-minute pump and was painted yellow and white. In 1993, LSI Fire rebuilt the Hahn to include a new body with additional compartments. The photograph above shows the Hahn as it looked prior to the 1993 rebuild. The photograph below shows the Hahn after it was rehabbed by LSI with additional compartments, front-hose intake, and updated warning lights. Still owned by Ashburn, this truck is a favorite among many members and is used as a ceremonial fire truck.

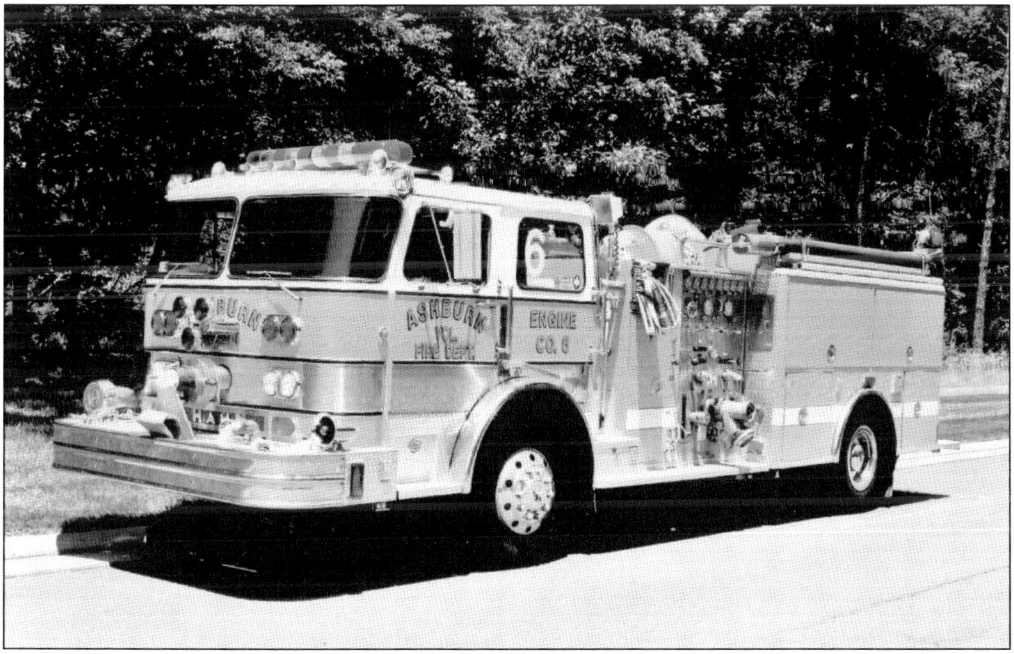

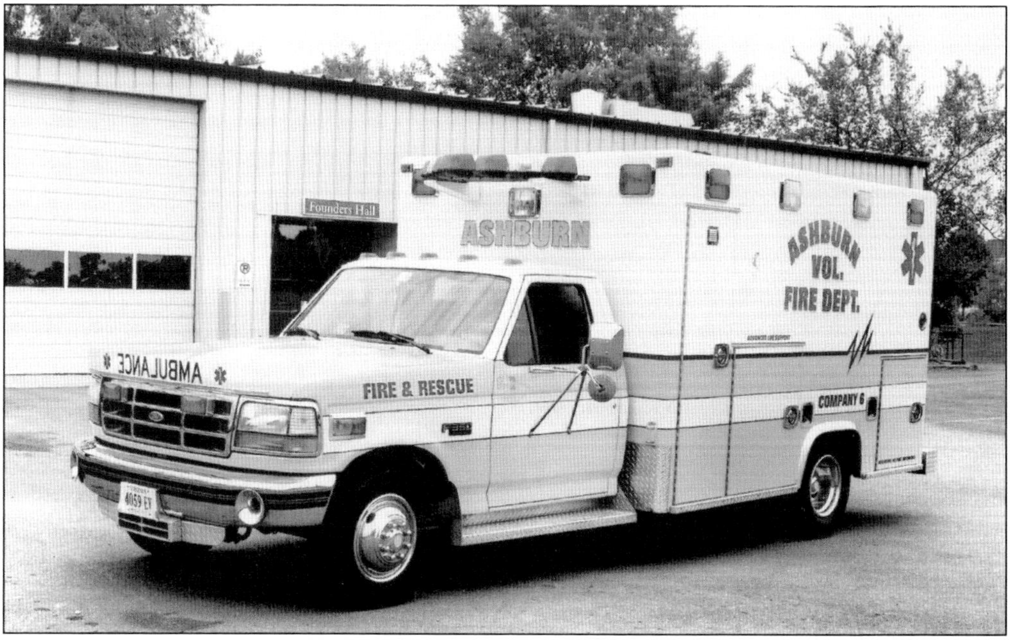

The year 1992 was an important one for Ashburn—it's when the department began to provide ambulance service to the community. Ashburn's first ambulance was a 1992 Ford F-350 built by Excellance, located in Alabama. Since 1992, the department has purchased five similar ambulances built by Excellance. The exception was in 1998, when the department placed into service a "monster medic," a larger ambulance built by Horton on a Freightliner chassis. The Freightliner was later sold. The photograph above shows one of the Ford F-350/Excellance ambulances, a 1994 model. The photograph below is Ambulance No. 6-3, Ashburn's monster medic. In 1996, to reflect the ambulance service now provided, the Ashburn Volunteer Fire Department was renamed the Ashburn Volunteer Fire and Rescue Department.

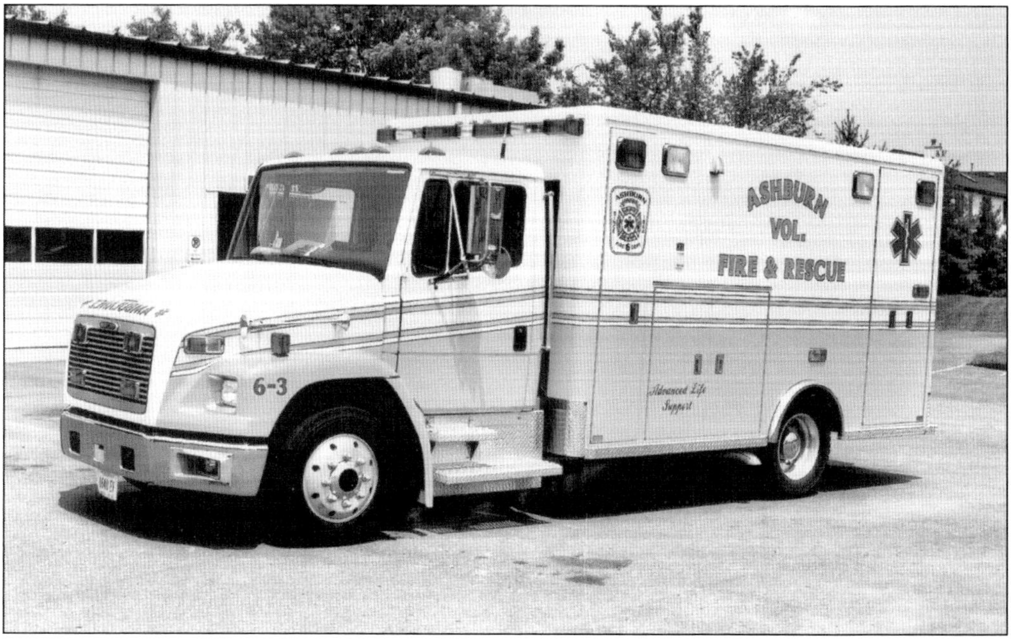

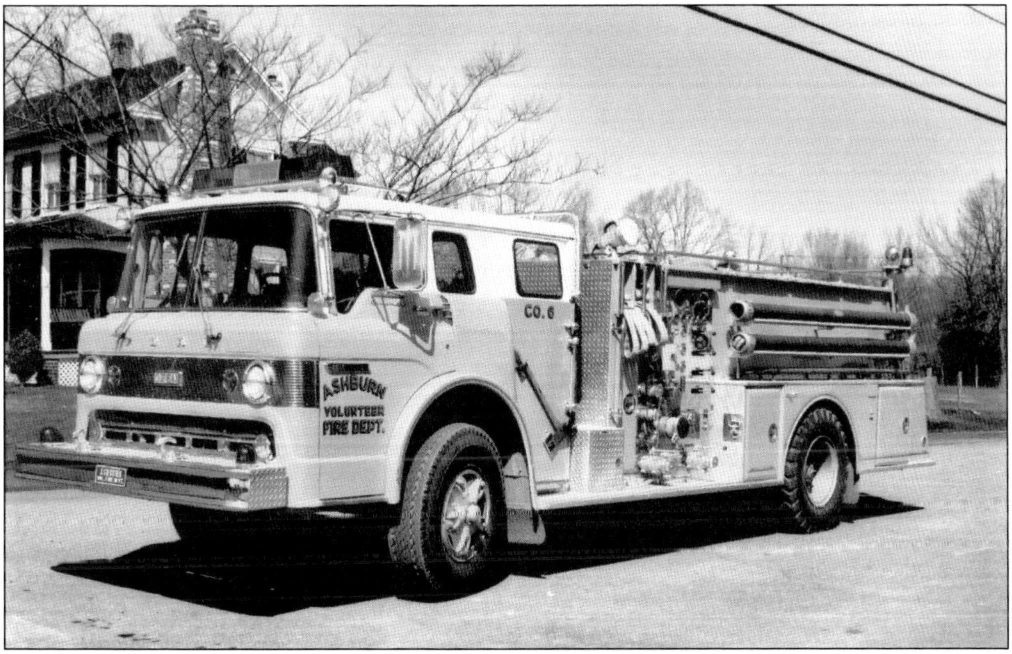

Ashburn's second fire truck built by Oren, of Roanoke, Virginia, was delivered on a 1977 Ford. Painted yellow and white, it had a 1,000-gallon-per-minute pump and held 600 gallons of water. Located directly behind the driver and officer and facing backwards was a bench for additional firefighters to sit and ride. The Ford C/Oren was later sold to a fire-apparatus dealer in Alabama.

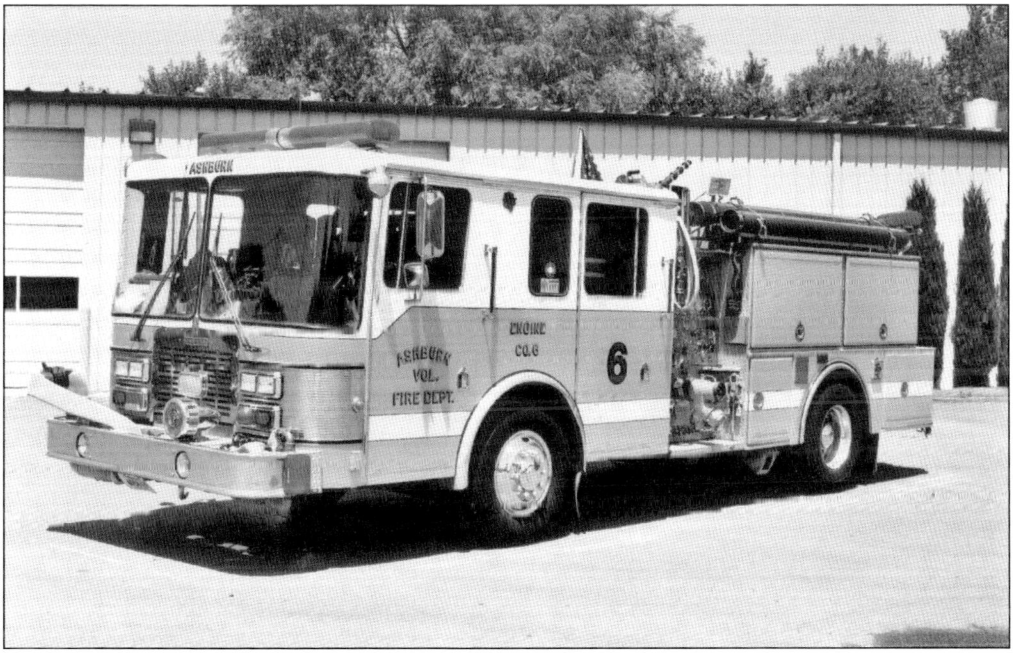

The first fully enclosed four-door pumper for Ashburn was Wagon No. 6, a 1991 Grumman FireCat equipped with a 1,250-gallon-per-minute pump and a 600-gallon water tank. Grumman acquired Oren and continued building fire trucks from Roanoke, Virginia. Wagon No. 6 was retired from service after serving the Ashburn community for 13 years.

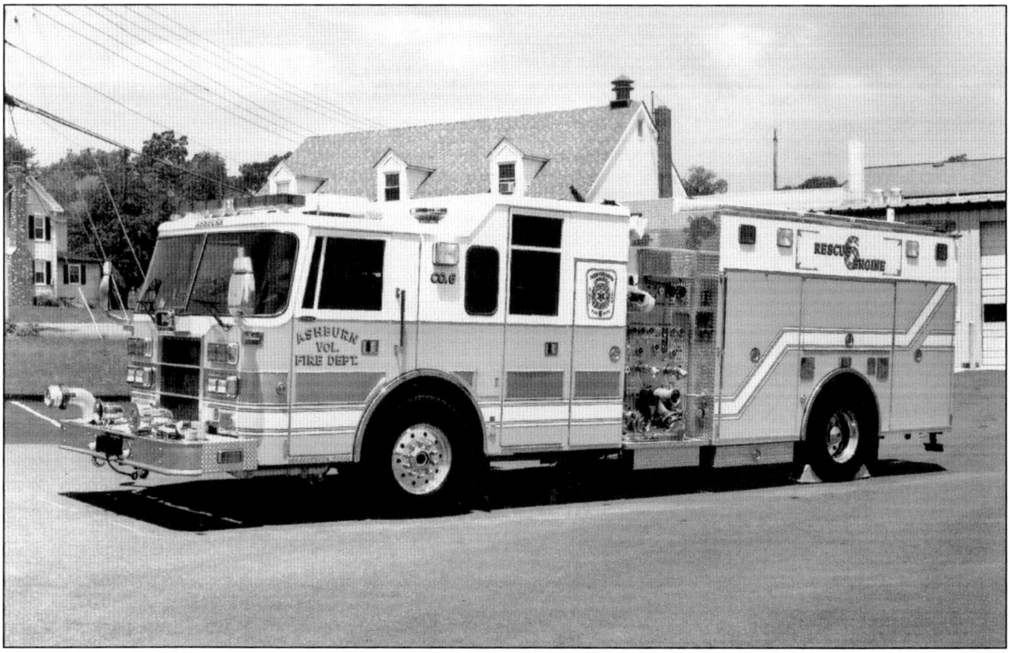

To carry more specialized gear required for automobile accidents and other situations, Rescue Engine No. 6 was placed into service in Ashburn during 1996. Built by Pierce Fire Apparatus of Appleton, Wisconsin, Rescue Engine No. 6 is a Lance model complete with a 1,250-gallon-per-minute pump, 500-gallon water tank, and two "jaws of life."

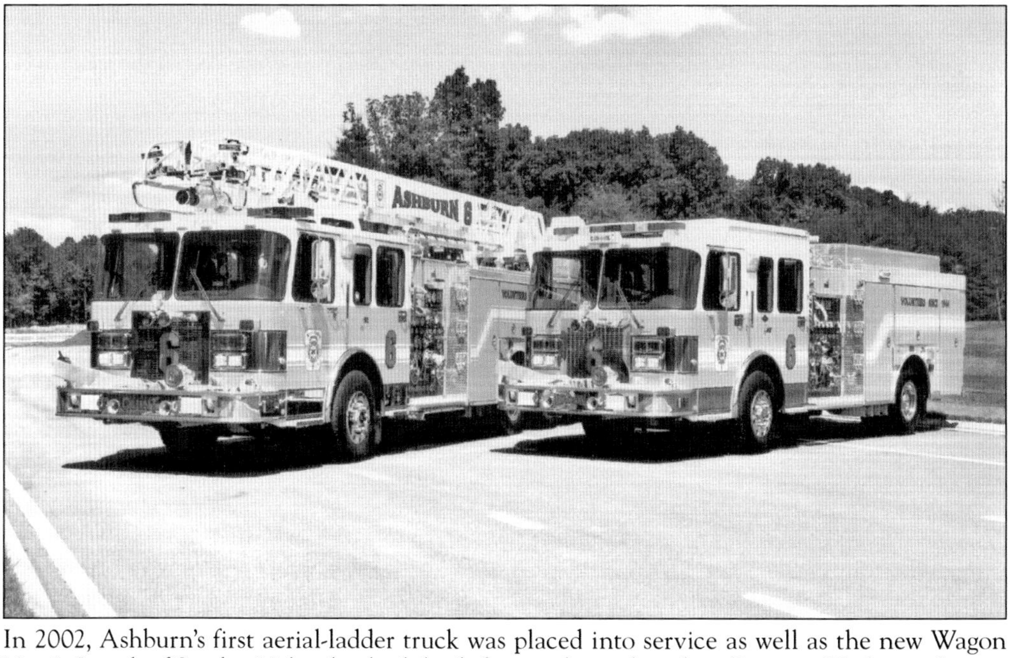

In 2002, Ashburn's first aerial-ladder truck was placed into service as well as the new Wagon No. 6. Smeal, of Snyder, Nebraska, built both fire trucks on four-door Spartan Gladiator chassis. Both trucks have 1,500-gallon-per-minute pumps. Wagon No. 6 holds 600 gallons of water, and Quint No. 6 holds 500 gallons. Quint No. 6 is also equipped with a 75-foot ladder and can operate as an engine or ladder truck.

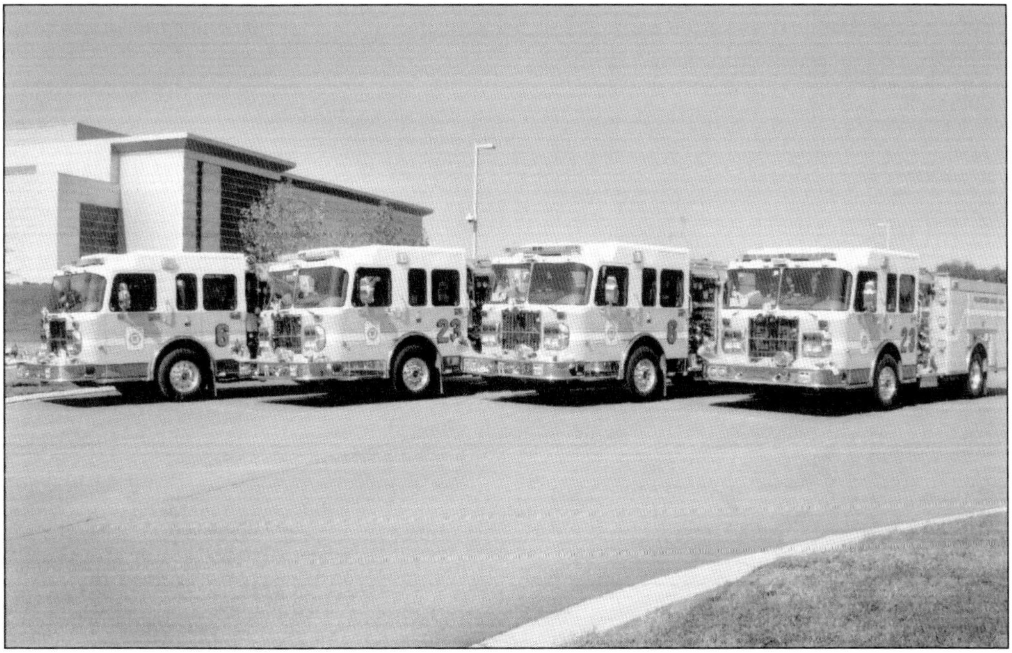

During 2003 and 2005, Ashburn accepted the delivery of four new fire engines. All trucks were built by Ferrara on Spartan Gladiator four-door cabs. The engines are equipped with 1,500-gallon-per-minute pumps and carry 500 gallons of water. Two of the engines were assigned to Station No. 6 (Ashburn) and two to Station No. 23 (Moorefield).

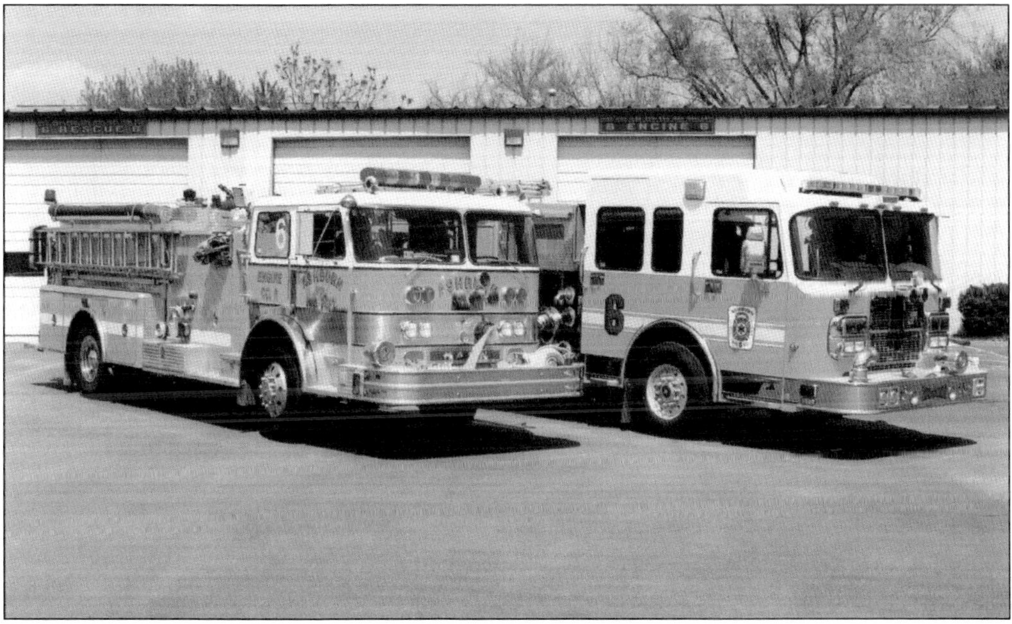

The delivery of the Spartan Ferrara engines during 2003 and 2005 meant the retirement of several Ashburn apparatus. Just after entering service, the new Engine No. 6 was photographed with the old Engine No. 6 in front of Station No. 6. The new engine is a 2003 Spartan/Ferrara (right), and Ashburn's old Engine No. 6 is a 1980 Hahn (left), now used by the department as a ceremonial engine.

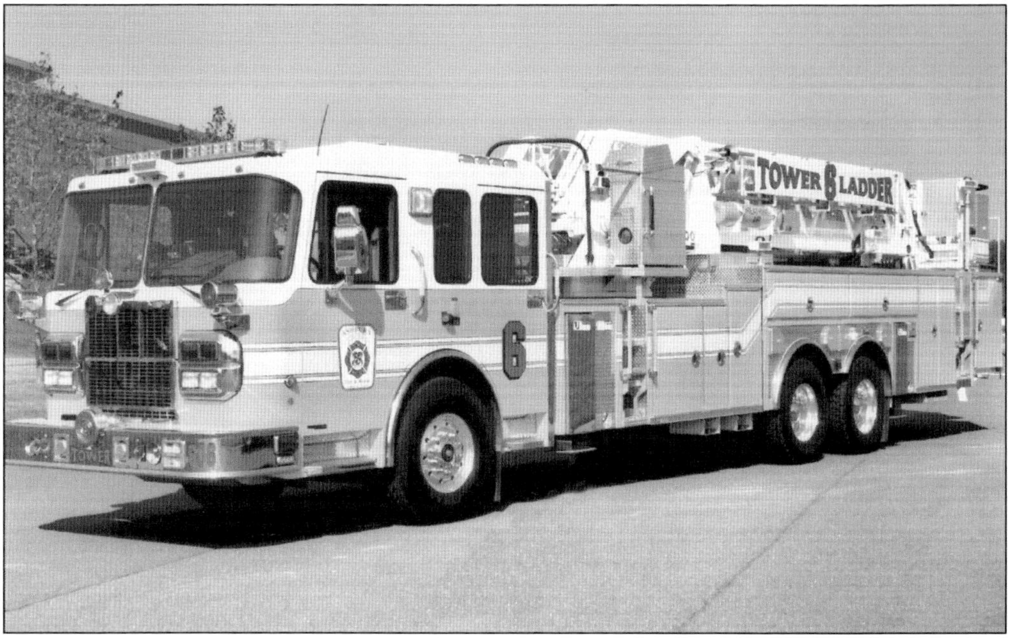

Tower No. 6 was proudly displayed in the 2006 calendar of Ferrara, the manufacturer who built the aerial tower for Ashburn. Built on a 2004 Spartan Gladiator chassis, Tower No. 6 has a 1,500-gallon-per-minute pump and a smaller 50-gallon water tank. A unique feature is that the pump panel is located on the cab roof so the operator is within easy reach of both the 100-foot aerial bucket controls and the pump panel controls.

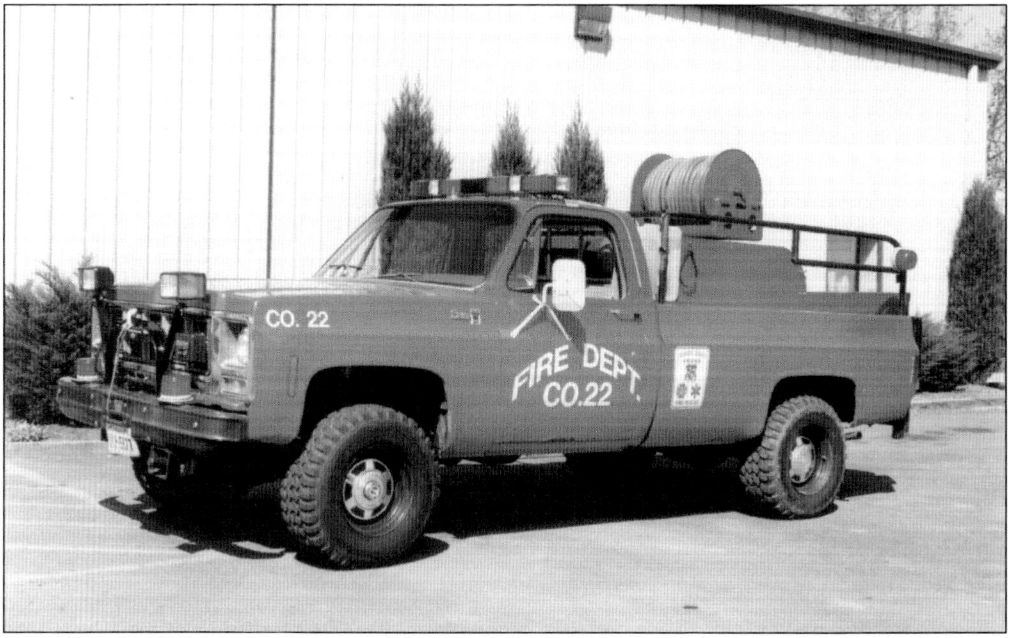

Xerox Corporation once operated a large complex in the Ashburn area and had a fire brigade on-site. The department was Station No. 22 in Loudoun County, and their roster included a brush unit and a utility truck. Brush No. 22 was a 1979 GMC red pickup truck. It often worked in conjunction with Ashburn units on emergency calls.

Two

Loudoun County Growth

Station Nos. 7–25

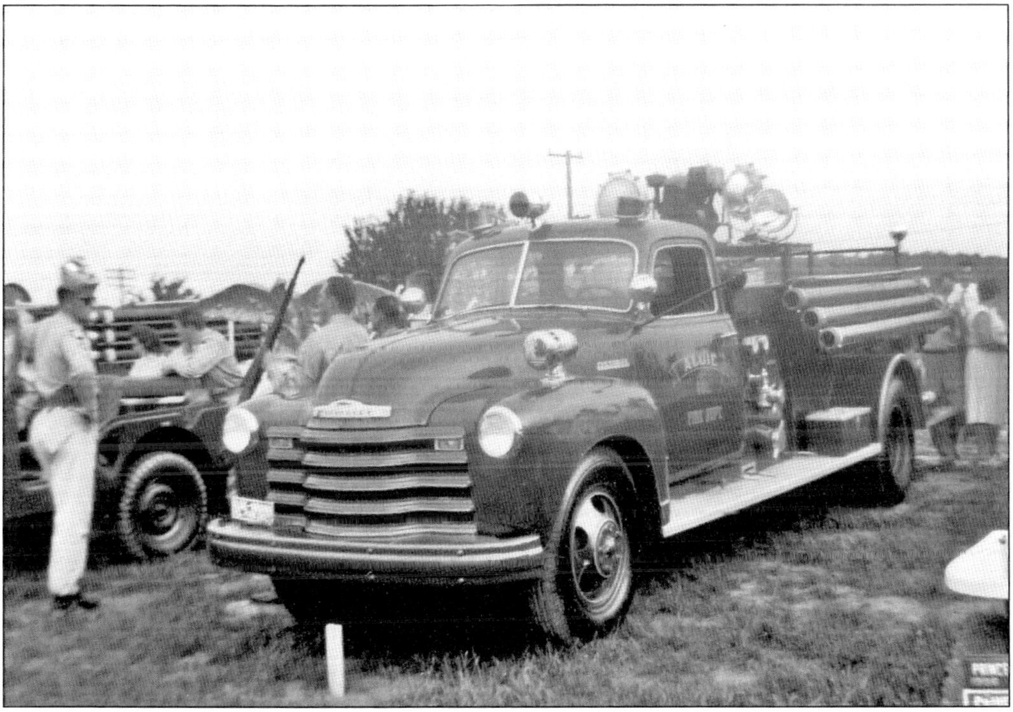

Aldie's first fire truck was purchased in 1955 when the department was established. Housed in the Jackson garage at the west end of the village, this 1948 Chevrolet was purchased used for $5,500 and had a 500-gallon-per-minute pump. The Chevrolet was totaled while parked in the fire station during the flooding caused by Hurricane Agnes in June 1972. (Courtesy of Pete West.)

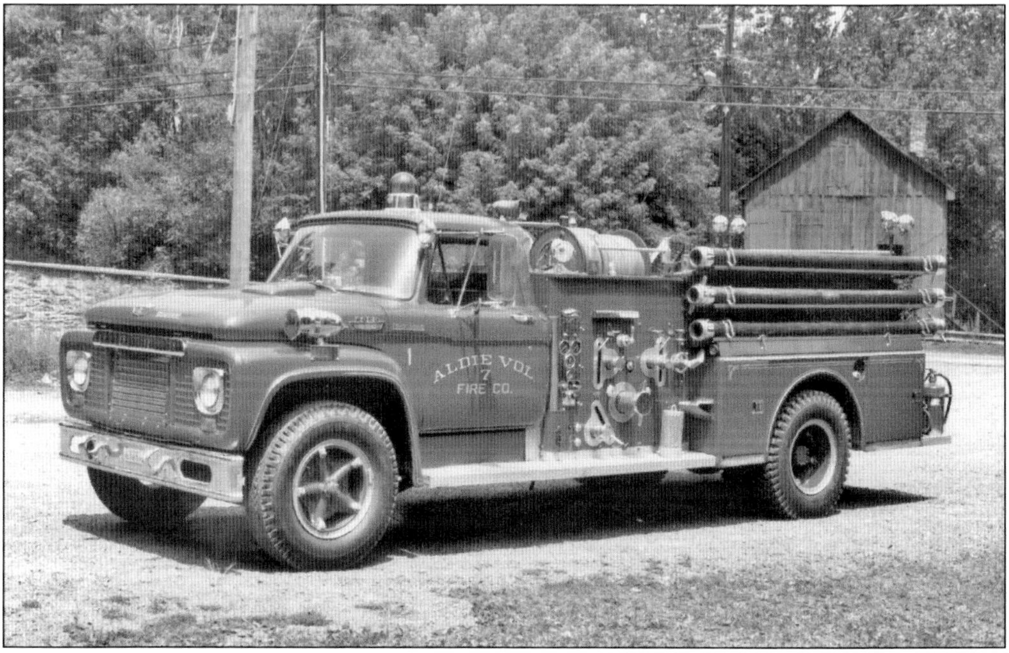

Aldie's first new fire truck was this 1963 Ford F-850, built by American LaFrance, of Emira, New York (a separate apparatus manufacturer from Ward LaFrance). Equipped with a 750-gallon-per-minute pump, it was purchased for $17,300. This Ford also suffered damage from flooding caused by Hurricane Agnes in June 1972 but was returned to service. (Courtesy of Chuck Madderom.)

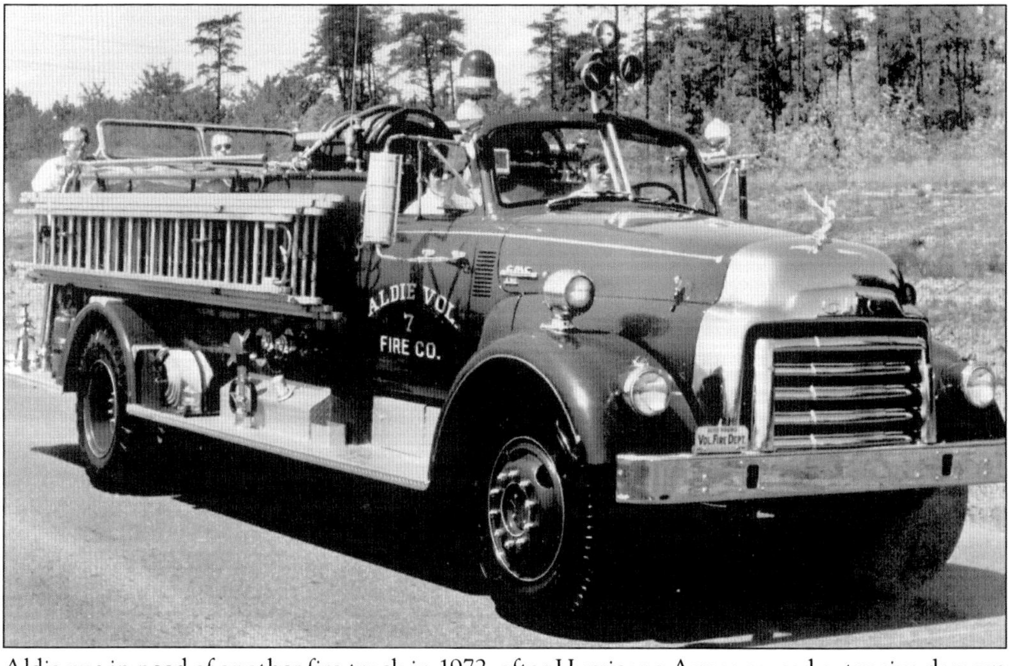

Aldie was in need of another fire truck in 1972, after Hurricane Agnes caused extensive damage to their 1948 Chevrolet. This 1953 GMC/American open-cab, 750-gallon-per-minute model was purchased from nearby Herndon, Virginia, for $3,500. Today a collector in the Northern Virginia area privately owns the GMC/American. (Courtesy of Aldie Volunteer Fire Department.)

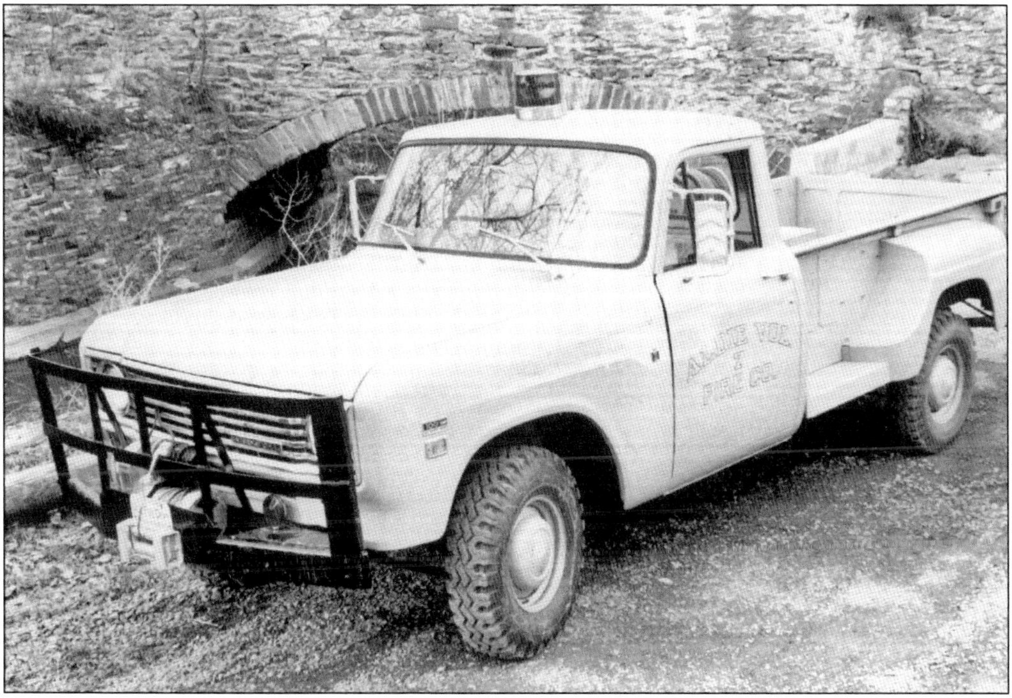

In 1974, the Aldie volunteers purchased a new International half-ton, four-wheel-drive pickup truck and had it outfitted as a brush truck. It was originally painted a lime-green color and equipped with a 90-gallon water tank. It was later repainted red and white and eventually was sold to a member of the Aldie Volunteer Fire Department. (Courtesy of Aldie Volunteer Fire Department.)

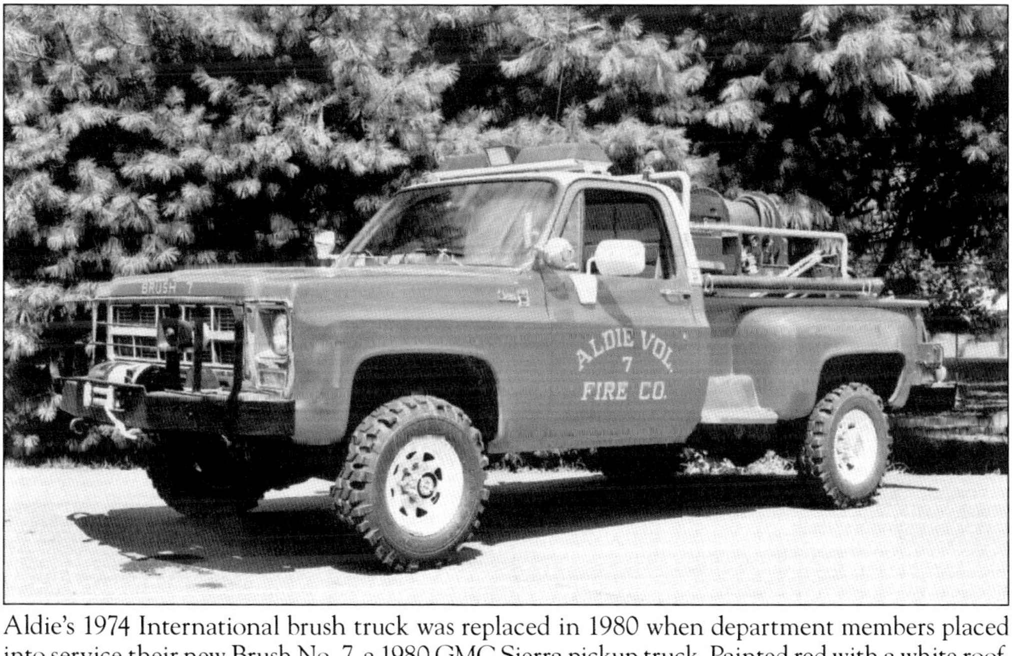

Aldie's 1974 International brush truck was replaced in 1980 when department members placed into service their new Brush No. 7, a 1980 GMC Sierra pickup truck. Painted red with a white roof, Brush No. 7 is equipped with a 250-gallon-per-minute pump and carries 200 gallons of water.

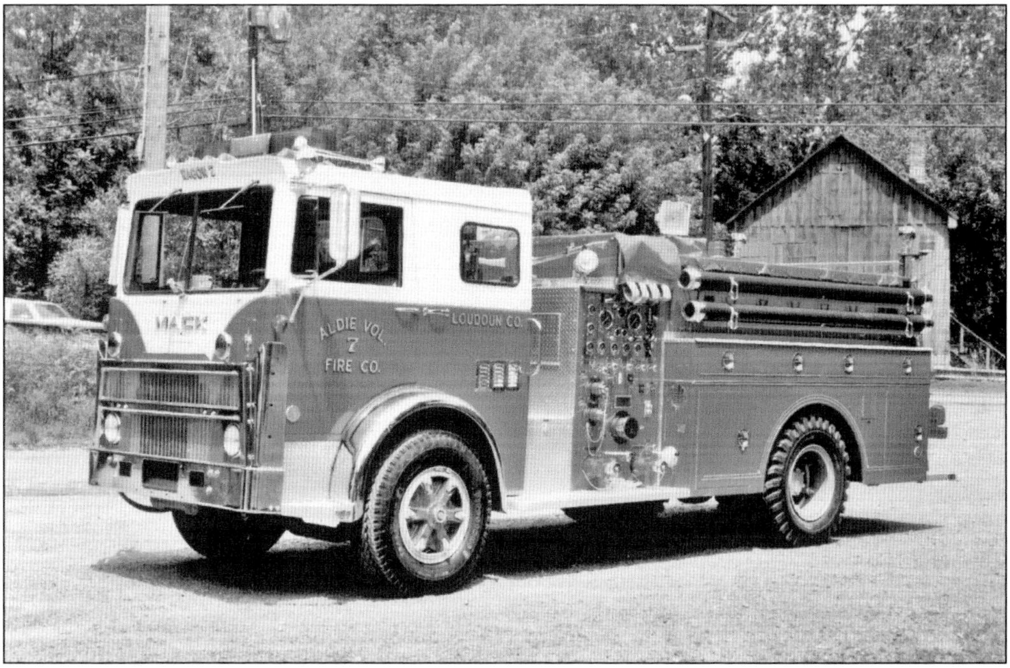

In 1979, for the first time, the Aldie Volunteer Fire Department responded to over 100 calls in a year for assistance. Also in 1979, with the help of generous donations, dinners, and other fund-raising projects, the fire department was able to purchase two new Mack fire trucks, each of which was equipped with a 750-gallon-per-minute pump. The Mack MB model (above) was nicknamed "Puppy," while the Mack MC model (below) was "Puppy II." Since Aldie's two new Macks were five-man canopy-cab fire trucks, all firefighters could be safely seated on the trucks rather than riding the rear back step. Puppy was sold in 1996 to an apparatus dealer in Alabama, and Puppy II was sold in 2005 to a department in Oklahoma. (Above, courtesy of Chuck Madderom.)

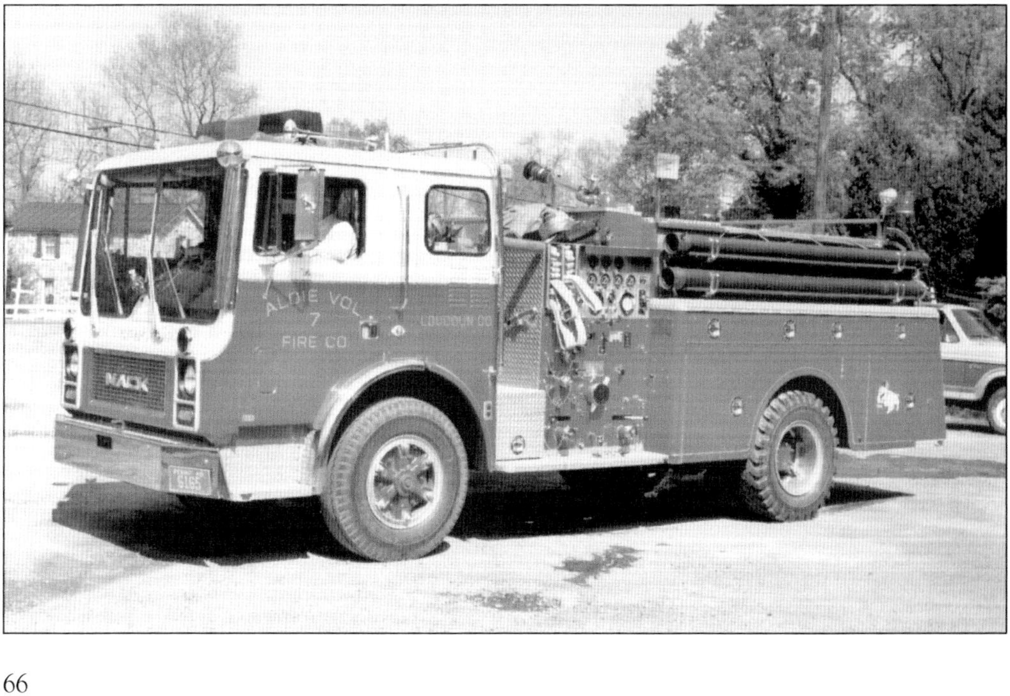

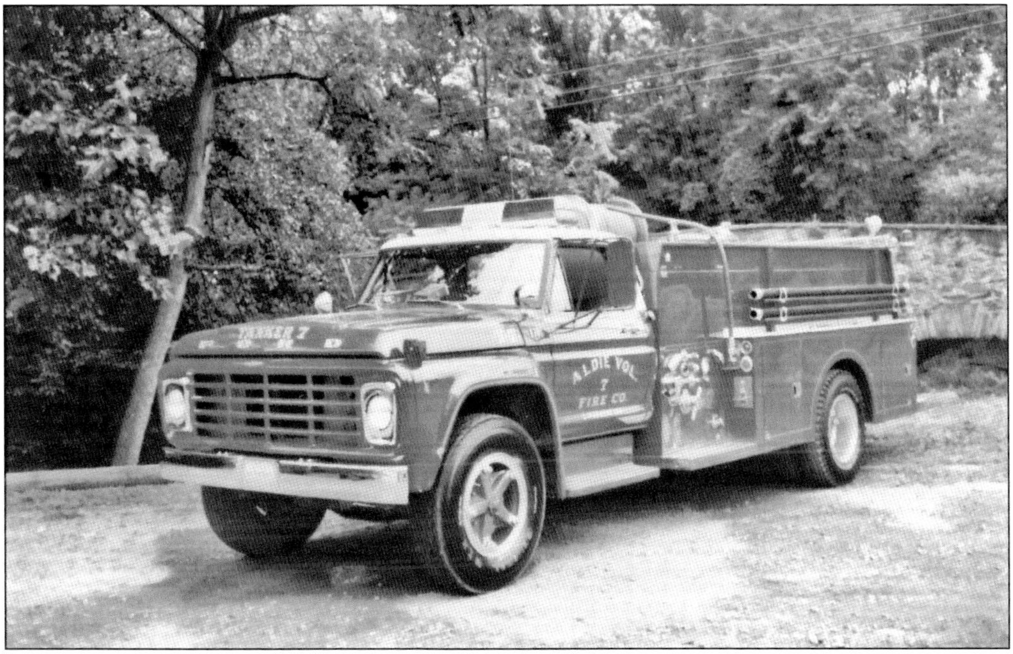

In 1974, a Ford F-700 tanker truck was purchased by Aldie and equipped by the Slagle's fire equipment company, located in South Boston, Virginia. This tanker replaced the last of a number of homemade tankers used by Aldie. Equipped with a booster hose reel on the top of the truck, Tanker No. 7 had a small 250-gallon-per-minute pump and carried 1,250 gallons of water. (Courtesy of Aldie Volunteer Fire Department.)

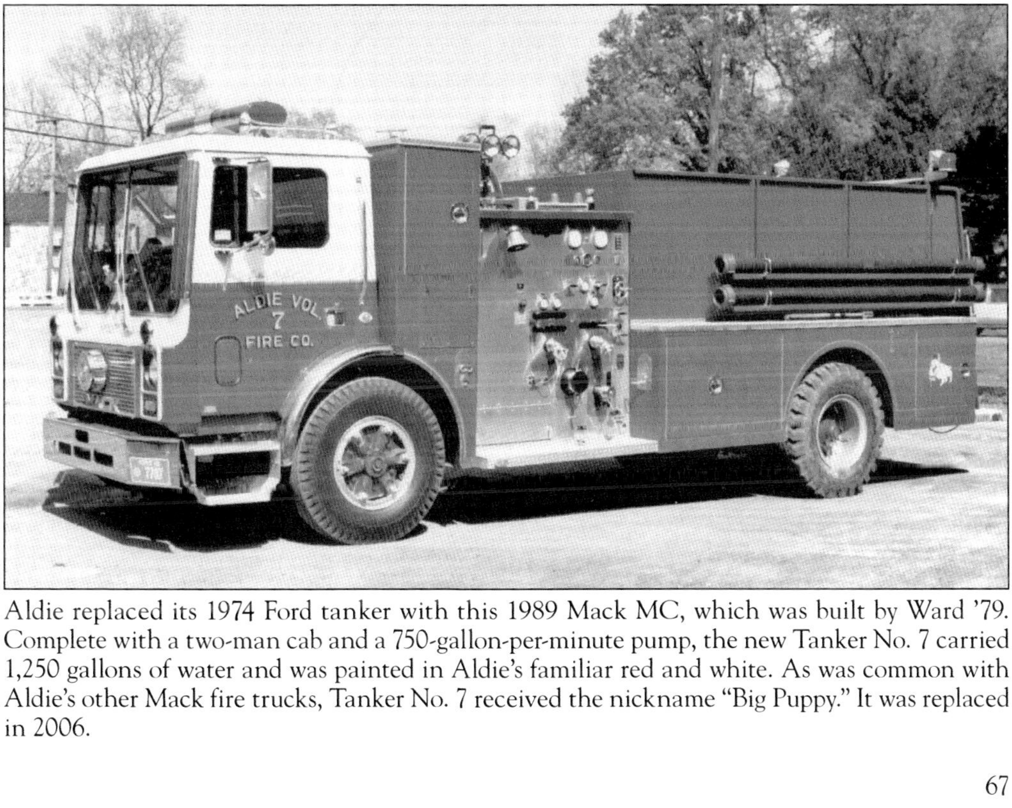

Aldie replaced its 1974 Ford tanker with this 1989 Mack MC, which was built by Ward '79. Complete with a two-man cab and a 750-gallon-per-minute pump, the new Tanker No. 7 carried 1,250 gallons of water and was painted in Aldie's familiar red and white. As was common with Aldie's other Mack fire trucks, Tanker No. 7 received the nickname "Big Puppy." It was replaced in 2006.

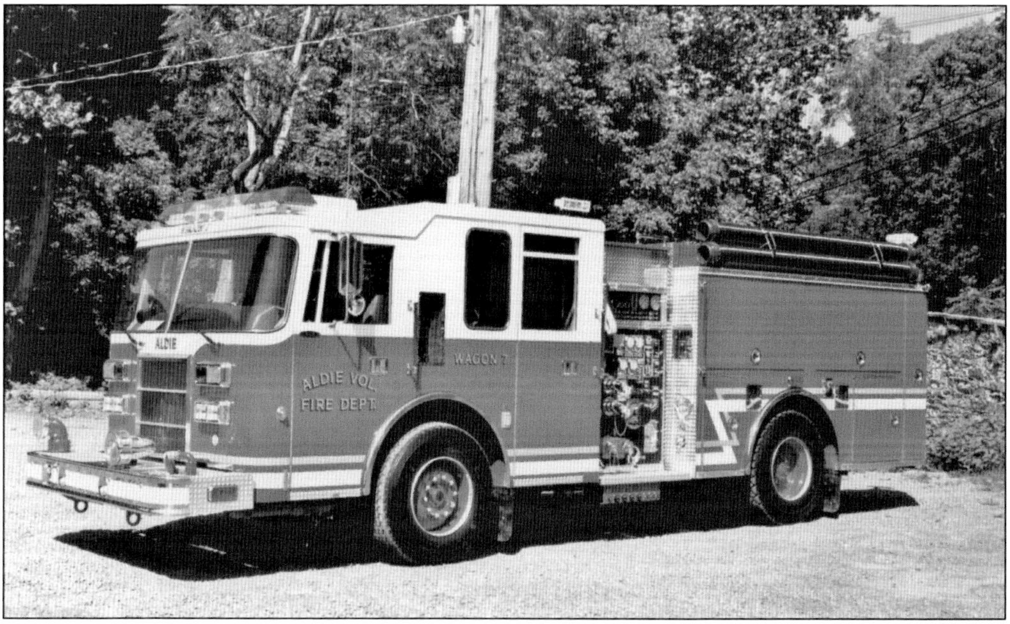

During the 1990s, Pierce Manufacturing of Appleton, Wisconsin, was delivering numerous fire trucks to Northern Virginia stations. The Aldie Volunteer Fire Department placed a new Wagon No. 7 in service during 1996, and it was a Pierce Dash. Wagon No. 7 was equipped with a 1,500-gallon-per-minute pump and carried 750 gallons of water. This Pierce also represented the first fully enclosed cab for the Aldie Volunteer Fire Department.

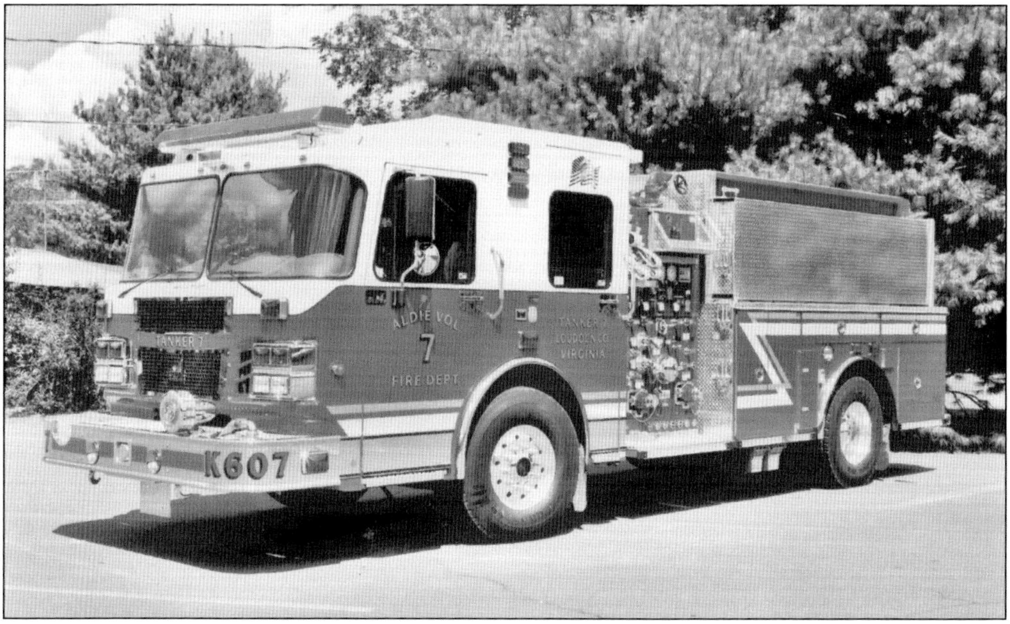

The Aldie community celebrated 50 years of fire service protection in 2005. That same year, a new tanker was placed into service. Built by Smeal with a Smeal Sirius cab, the tanker included a four-door cab, a 1,500-gallon-per-minute pump, 1,760 gallons of water, 40 gallons of foam, supply hose, attack hose lines, and compartments to carry other necessary equipment. Tanker No. 7 replaced the 1989 Mack tanker.

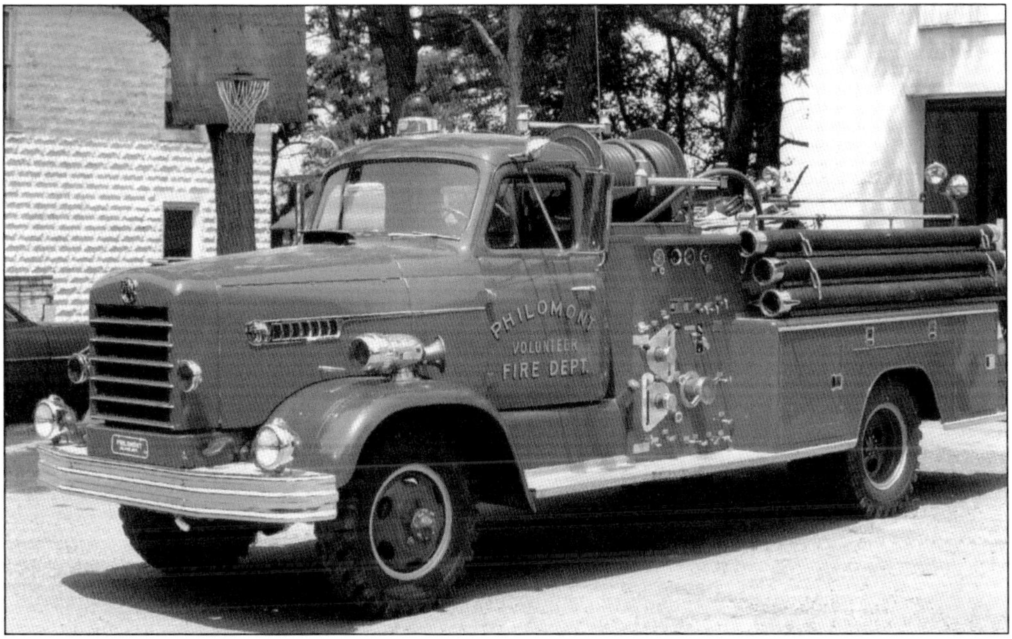

Several fires during 1955 in the Philomont area prompted the community to establish a volunteer fire department. A local resident, Capt. Richard Patch, provided an incentive. If the Philomont community would build a fire station, Captain Patch would purchase a fire truck. The station was built, and a year later, a 1956 Four Wheel Drive Company (FWD) fire truck was delivered to Philomont. This 1956 FWD is currently under restoration. (Courtesy of Robert Kimball.)

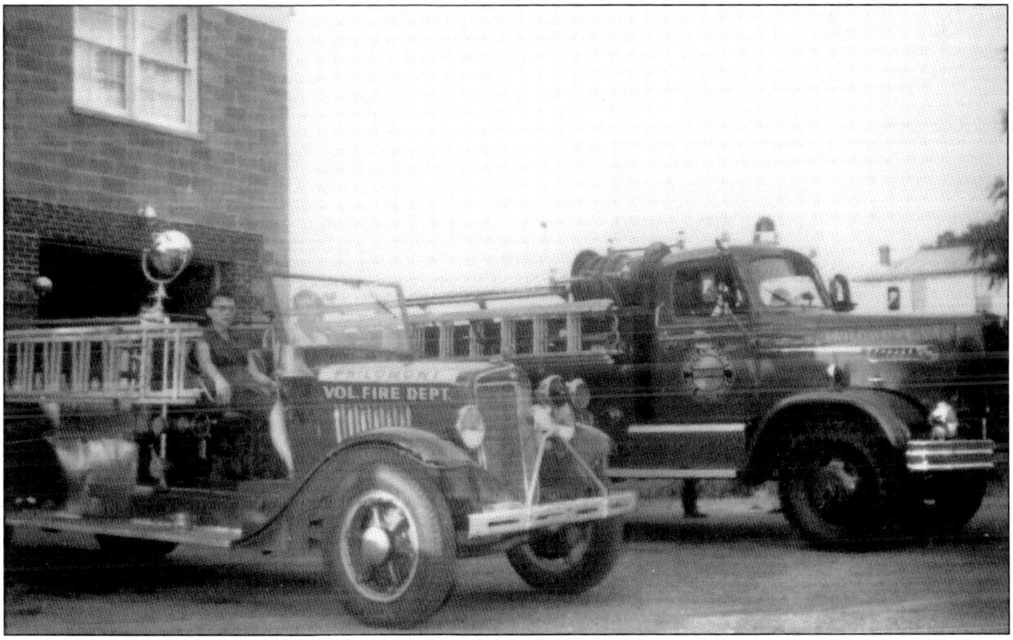

Philomont purchased a 1935 International fire truck in the late 1950s. This International was purchased from a Maryland fire department and worked in conjunction with the 1956 FWD. The photograph above was taken in June 1957 and shows both the International and FWD at Station No. 8. (Courtesy of Pete West.)

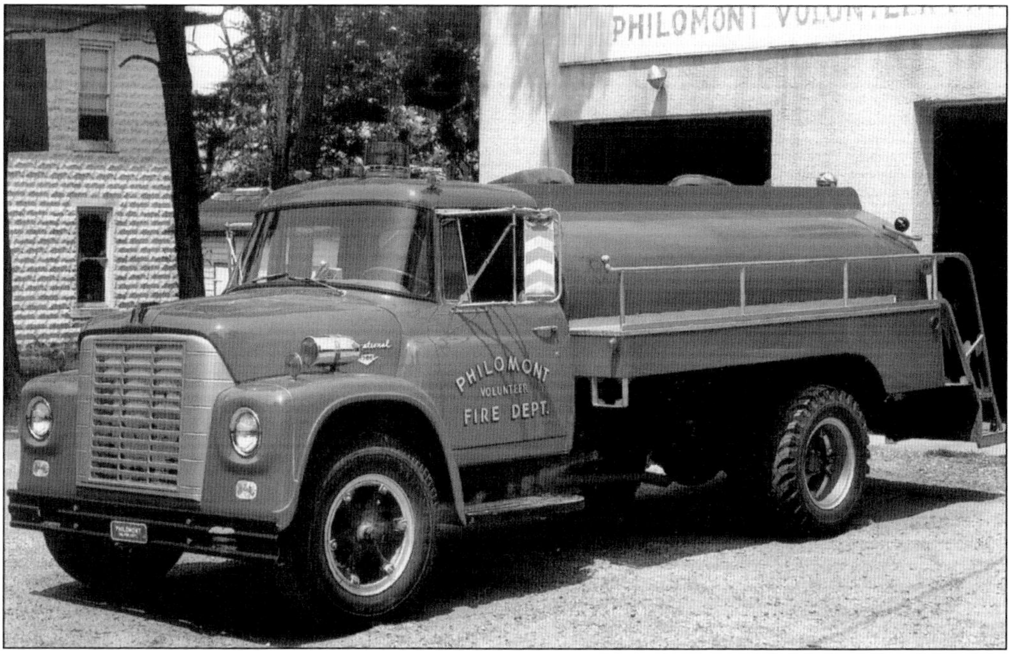

Lack of water can be a challenge in Philomont. The first tanker at Station No. 8 was purchased in the late 1950s and was an old GMC military tanker. In 1967, the 1,500-gallon tank was moved onto this new 1967 International. This tanker was upgraded again in 1980, when the 4-Guys fire truck company of Pennsylvania upgraded the tanker body. In 1996, Tanker No. 8 was sold to Station No. 16 in Neersville.

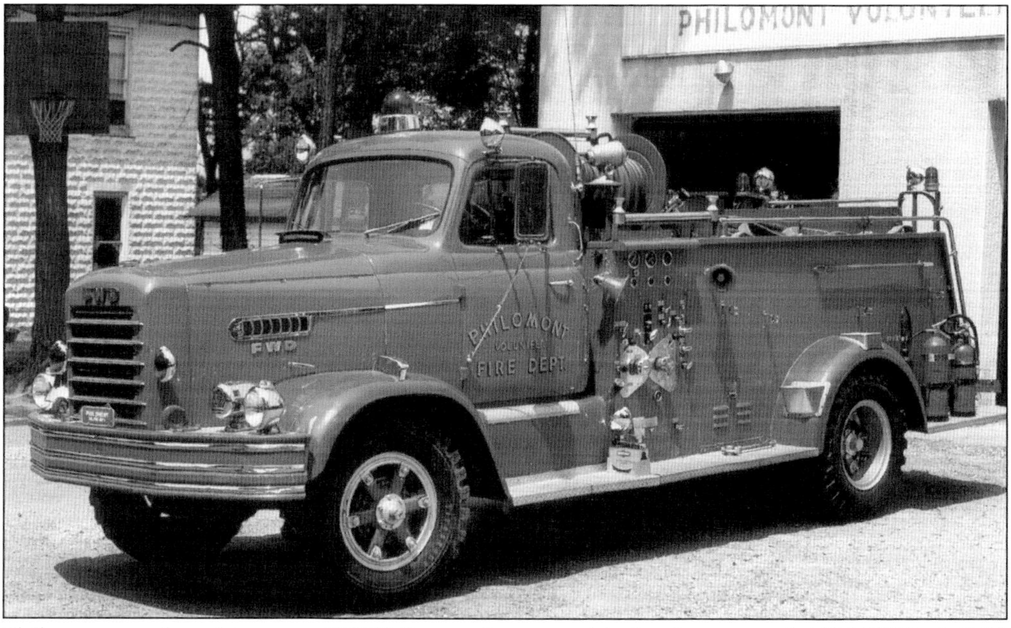

Philomont purchased its second FWD fire truck to run with the department's 1956 FWD. A 1962 model was purchased from the Broomall, Pennsylvania, fire department in 1974. Equipped with a 500-gallon-per-minute pump, this FWD was later sold by Philomont to the nearby Upperville Volunteer Fire Department in Fauquier County. (Courtesy of Robert Kimball.)

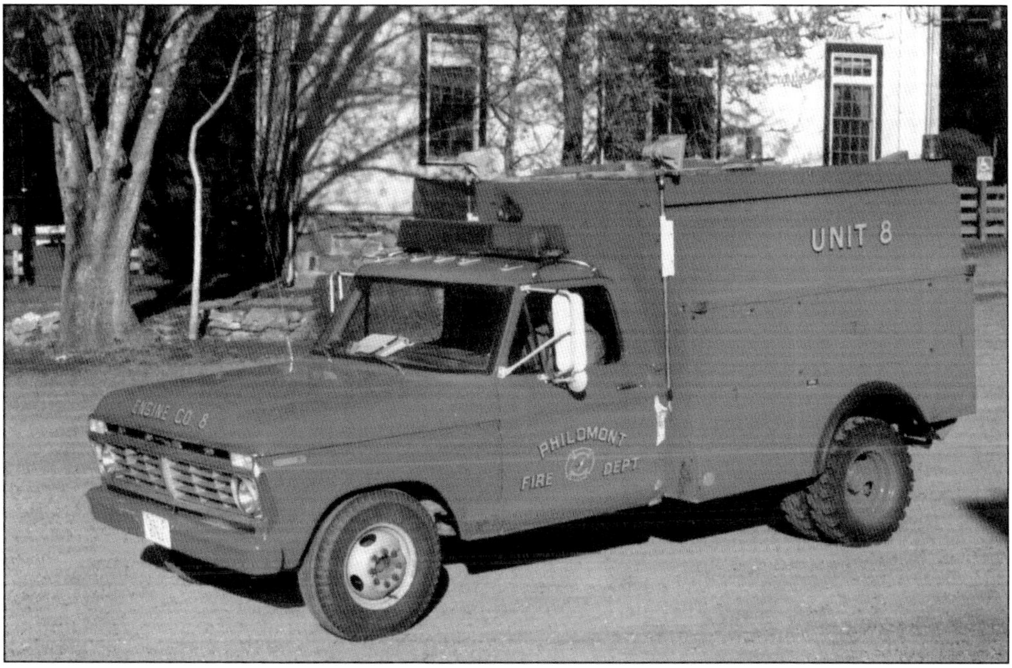

Small equipment trucks were common in Loudoun County fire stations, including Philomont. These trucks carried a variety of firefighting equipment to the emergency scene. The Bell Atlantic phone company donated this 1975 Ford/Baker truck to the volunteers at Philomont, and it was designated as Unit No. 8. The Ford was later sold to a farmer in the Philomont area. (Courtesy of Dave Bowen.)

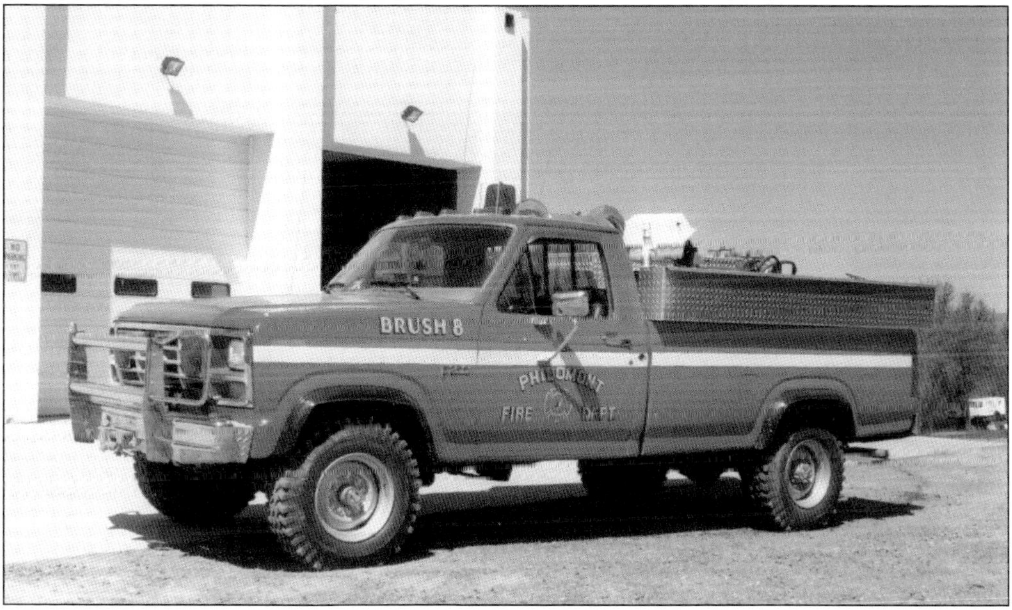

The 4-Guys fire truck company, located in Meyersdale, Pennsylvania, built this 1983 Ford F-250 brush truck for Philomont. 4-Guys is best known for its tanker trucks, so a smaller brush truck by this manufacturer was a bit uncommon. Philomont runs the Ford, equipped with a 250-gallon-per-minute pump and a 250-gallon water tank, as their Brush No. 8.

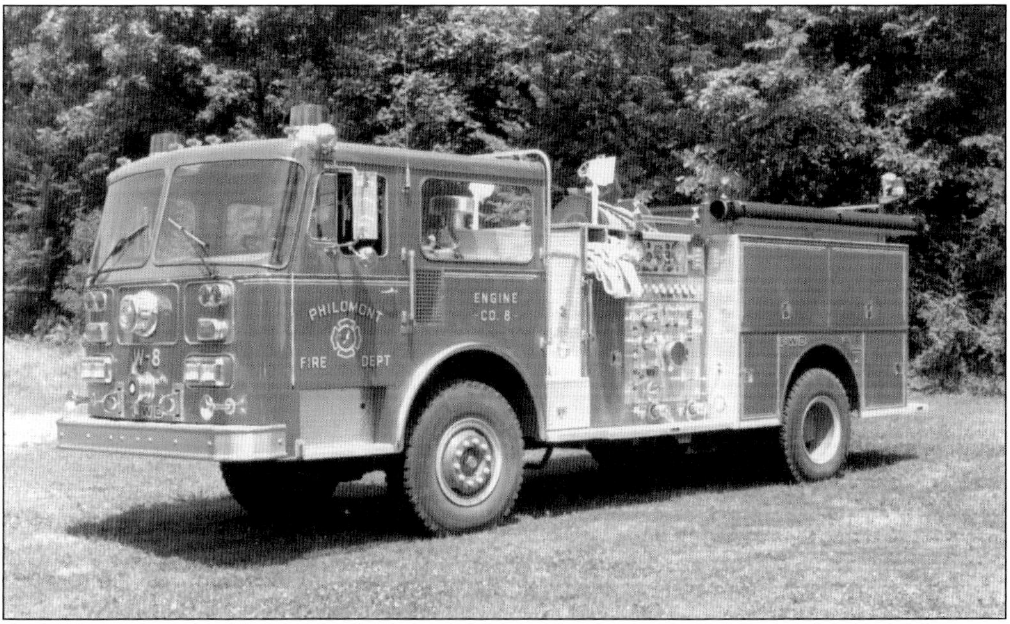

Philomont liked their earlier FWD fire trucks, and when it came time to order another fire truck, FWD was the choice. The department's third FWD was purchased new in 1987, and it was this 1,250-gallon-per-minute pumper equipped with a 750-gallon water tank. Originally known as Wagon No. 8, the FWD is capable of maneuvering on Philomont's smaller roads during inclement weather.

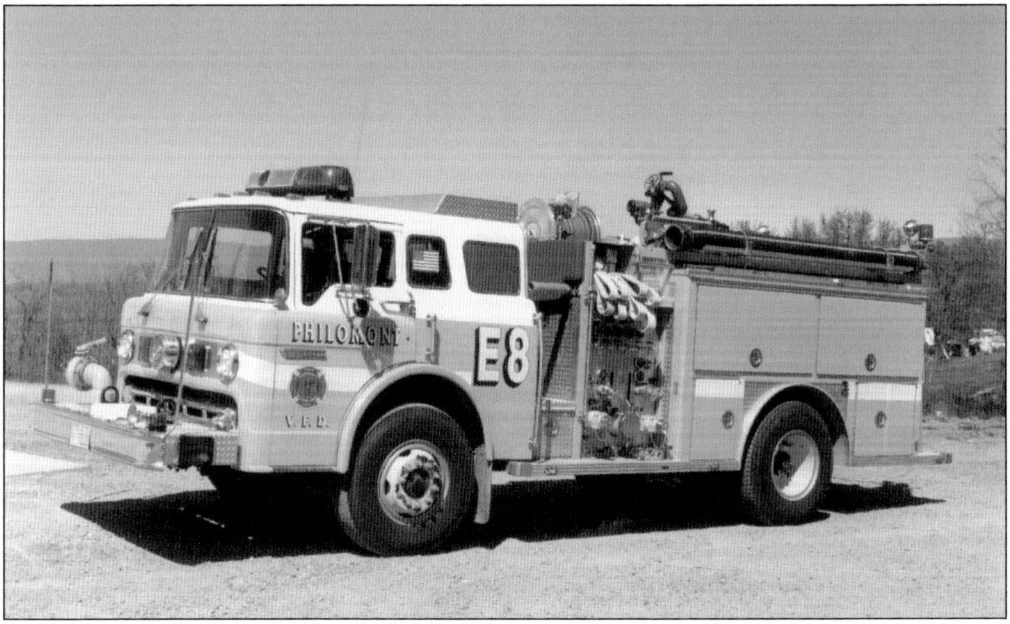

In 1991, Philomont purchased this 1985 Ford C fire truck built by E-One. Painted lime green and white, this truck originally was operated by the Burke Volunteer Fire Department in nearby Fairfax County. Engine No. 8 was equipped with a 1,000-gallon-per-minute pump and carried 500 gallons of water. Philomont never repainted the Ford, and it was later sold to a fire-apparatus broker in 2002.

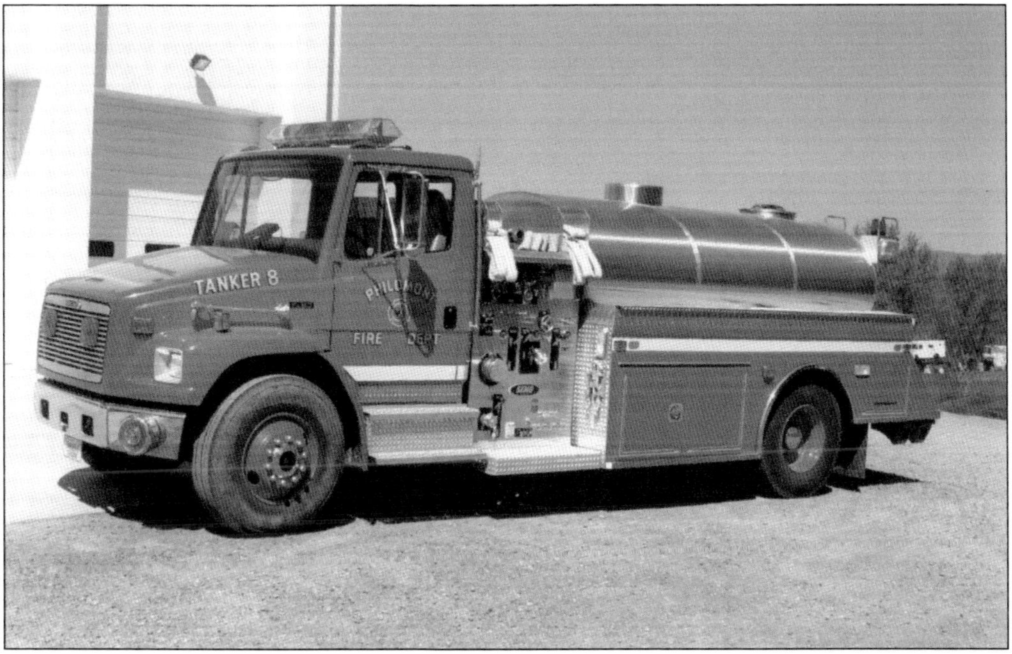

In 1996, Philomont's 1967 International tanker was sold to the Neersville Volunteer Fire Department in Loudoun County. Replacing the International tanker was this Freightliner 80 built by Semo. The new Tanker No. 8 was equipped with a 750-gallon-per-minute pump, an 1,850-gallon water tank, and several attack hose lines.

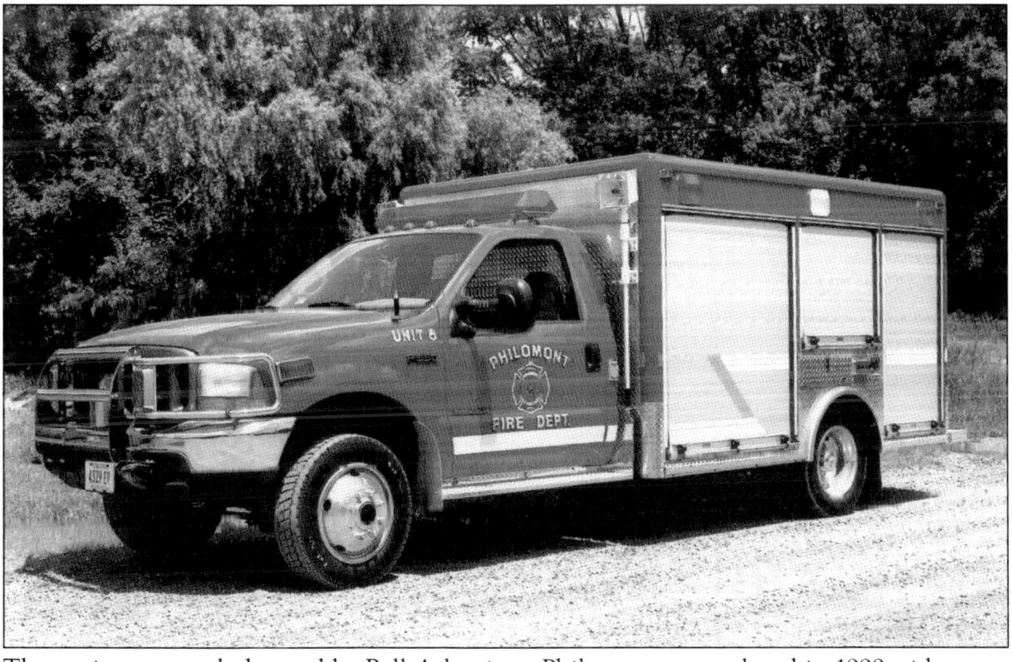

The equipment truck donated by Bell Atlantic to Philomont was replaced in 1999 with a new Unit No. 8. A 1999 Ford F-550 provides the power, and the rescue box with roll-up doors was built by LSI Fire. Unit No. 8 is equipped with a variety of tools and equipment, including a light tower, an air cascade system, and extrication tools.

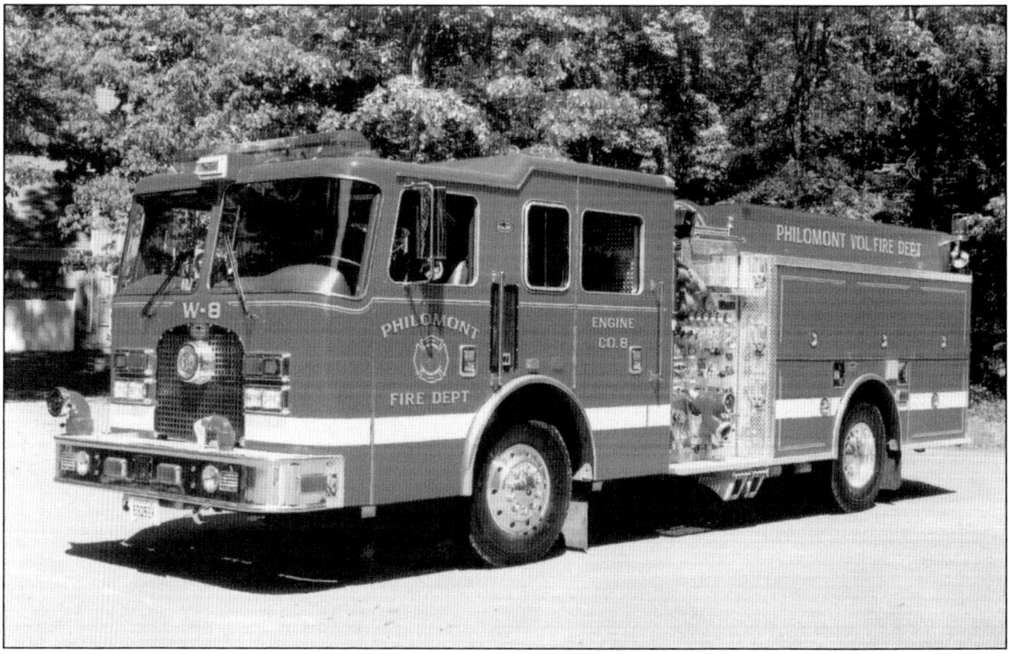

KME Fire Apparatus, a fire-truck builder in Pennsylvania, delivered Philomont's first four-door enclosed fire truck in 2002. Wagon No. 8 is equipped with a 1,250-gallon-per-minute pump, carries 750 gallons of water, and holds 40 gallons of foam. The KME was built with a raised-roof cab area in the back to give firefighters some additional room and comfort.

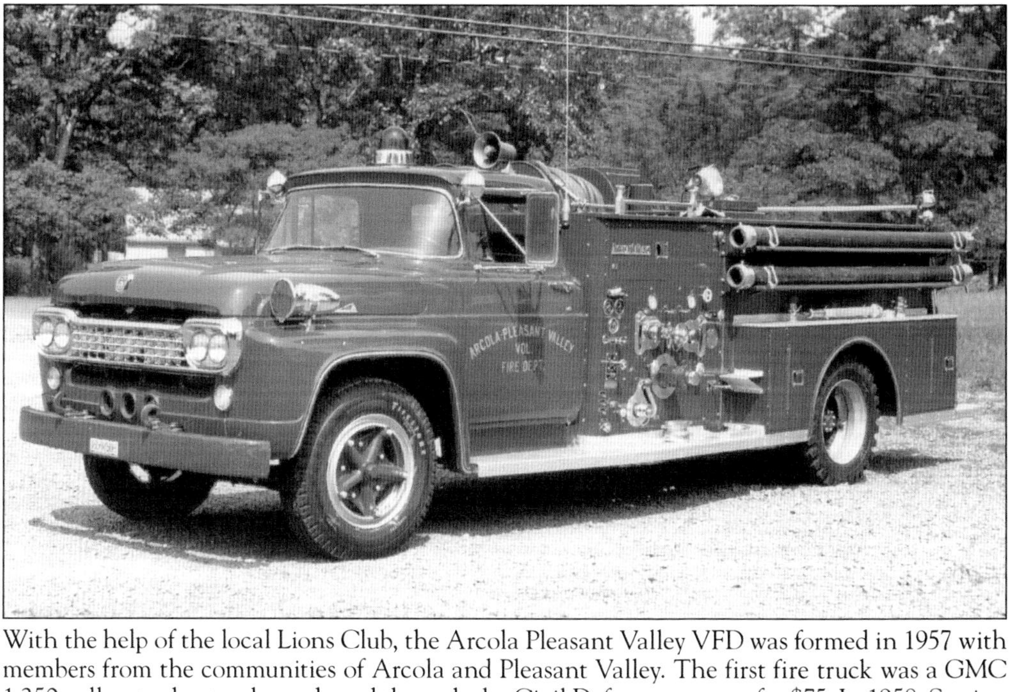

With the help of the local Lions Club, the Arcola Pleasant Valley VFD was formed in 1957 with members from the communities of Arcola and Pleasant Valley. The first fire truck was a GMC 1,250-gallon tanker truck purchased through the Civil Defense program for $75. In 1958, Station No. 9 purchased their second fire truck for $13,000 (above). This Ford was built by American LaFrance and was equipped with a 750-gallon-per-minute pump. (Courtesy of Robert Kimball.)

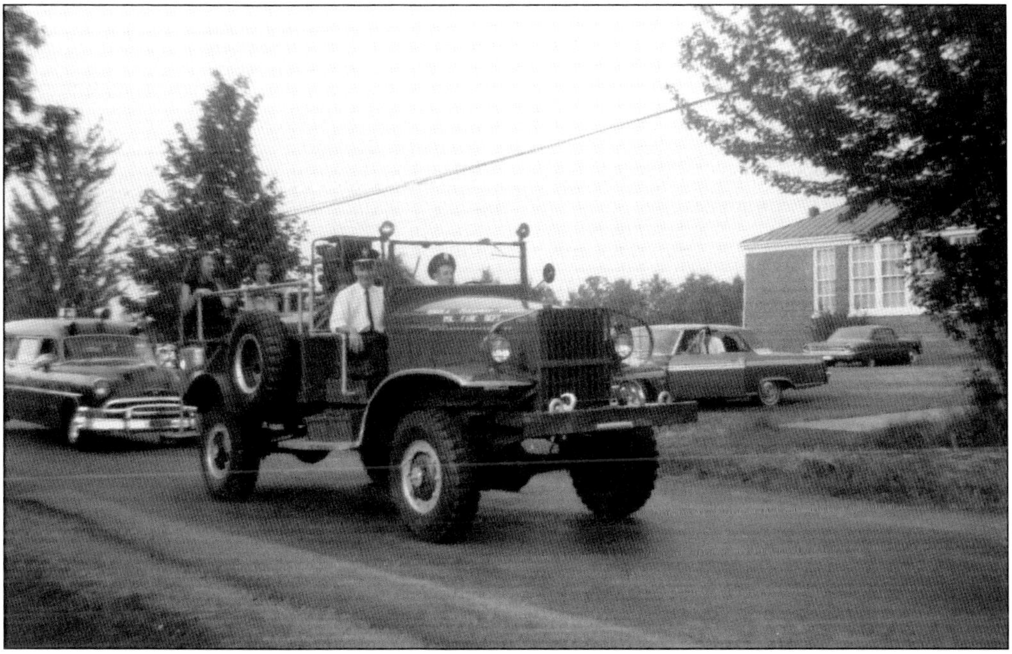

Arcola's third fire truck was obtained through the military and was this 1942 International that was previously a military weapons carrier. The Arcola volunteers equipped it as a brush truck. Note the open cab and large spare tire mounted on the side of the International. In 1971, Arcola sold this brush truck to Round Hill Station No. 4. (Courtesy of Pete West.)

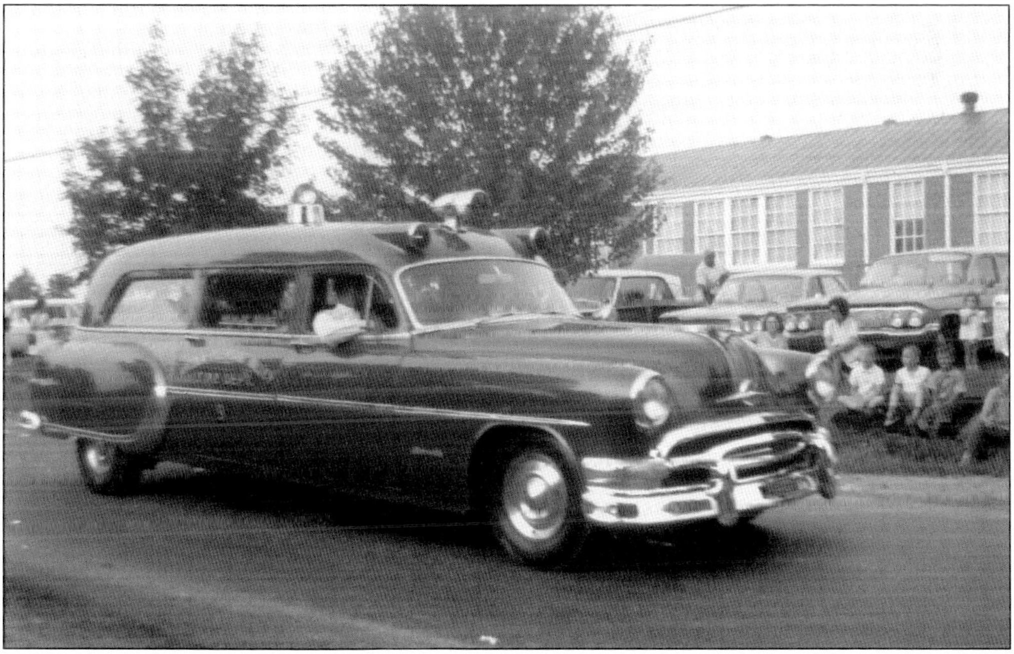

In 1960, Arcola placed its first ambulance into service. This Pontiac ambulance was purchased from the Bailey's Cross Roads Volunteer Fire Department in Fairfax County, Virginia. Arcola would later purchase a 1962 International Carryall and a 1969 Chevrolet Carryall for use as ambulances. (Courtesy of Pete West.)

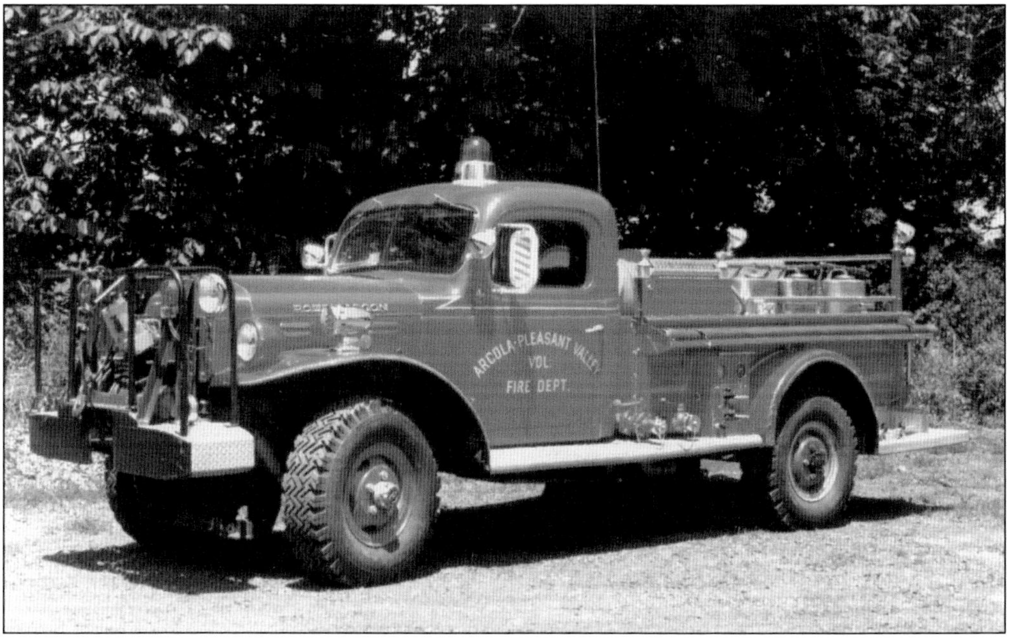

The Young Fire Equipment Corporation was a popular apparatus manufacturer for fire departments in Northern Virginia. Young built a variety of fire trucks, including smaller brush units. Arcola's Brush No. 9 is a 1969 Dodge Power Wagon with a 250-gallon-per-minute pump and a 250-gallon water tank. Although the Dodge is no longer in active service, it is still owned by Station No. 9, and plans are set for a complete restoration.

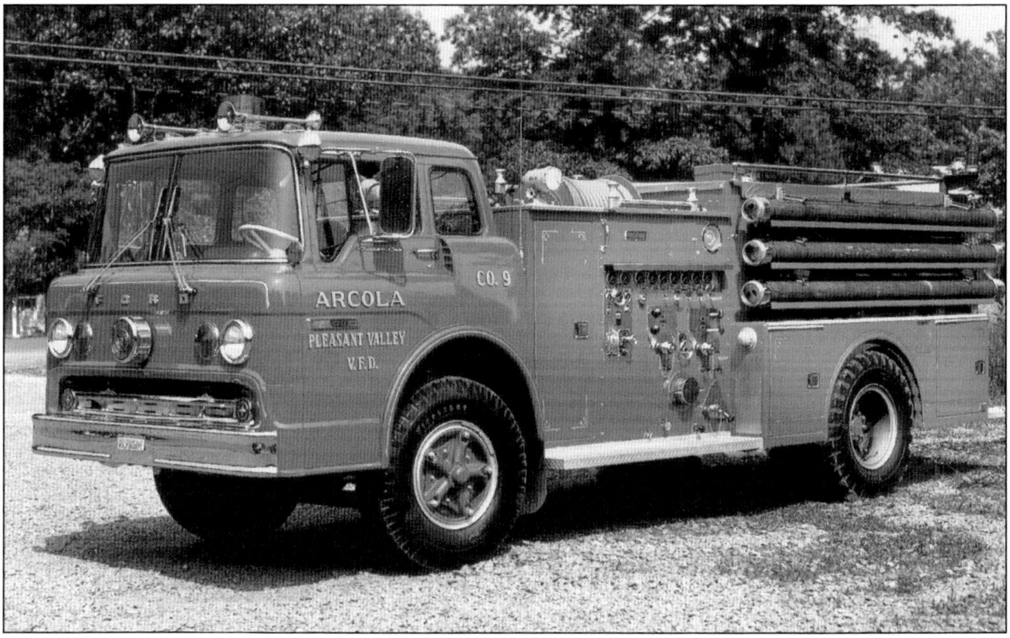

Although they had similar names, American LaFrance and Ward LaFrance were separate fire-apparatus manufacturers. Ward LaFrance delivered a variety of fire trucks to fire stations in Loudoun County. Arcola purchased this red 1971 Ford C/Ward LaFrance, which was equipped with a 750-gallon water tank. Note the two air horns on the cab roof. (Courtesy of Robert Kimball.)

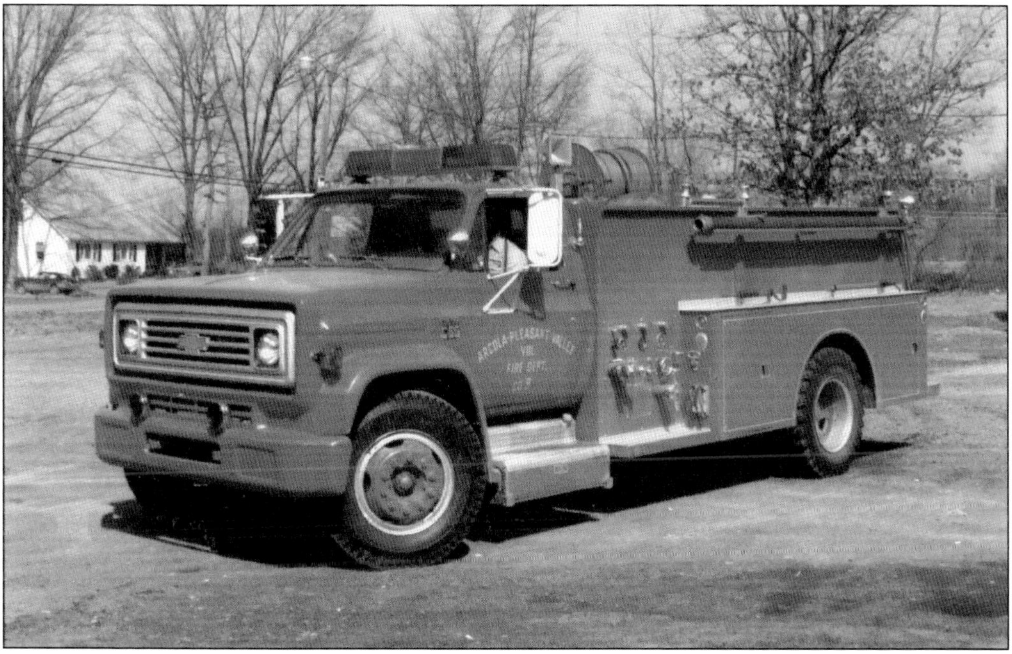

Arcola Station No. 9 volunteers placed a new tanker in service in 1973. Painted red, Tanker No. 9 was a Chevrolet with a tanker body built by Atlas. It was equipped with a small pump and held 1,200 gallons of water. Tanker No. 9 was sold in the early 1990s to a fire-truck distributor in West Virginia.

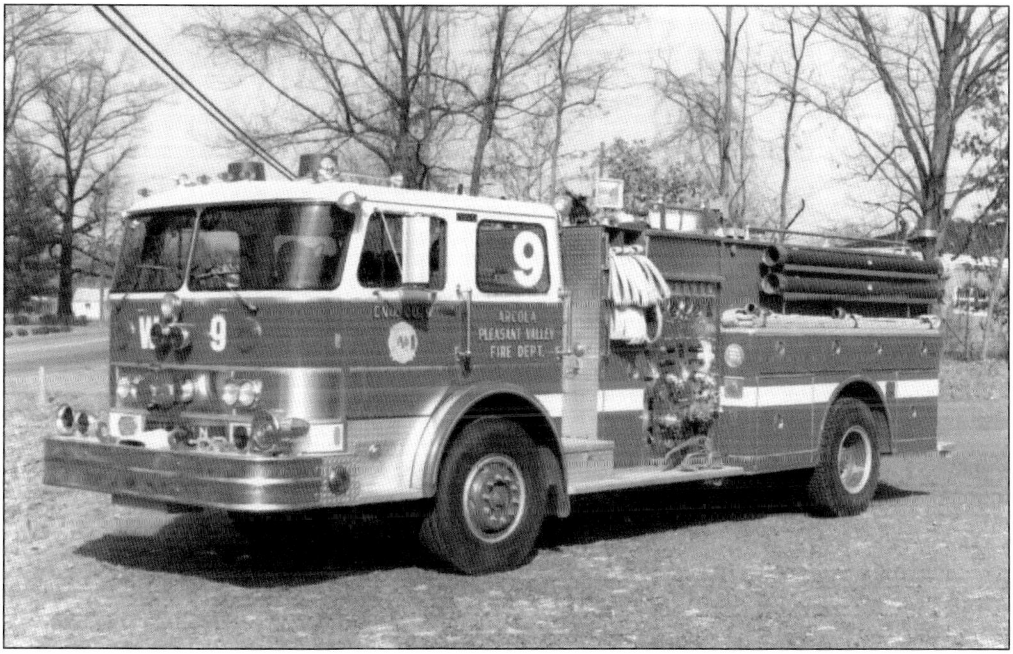

Equipped with two emergency warning lights on the cab roof and a Roto-Ray warning light on the front cab, Wagon No. 9 from Arcola operated a 1982 Hahn fire truck that held 750 gallons of water and was equipped with a 1,000-gallon-per-minute pump. The red-and-white Hahn was a favorite of many members at Arcola Pleasant Valley and was replaced in 2006.

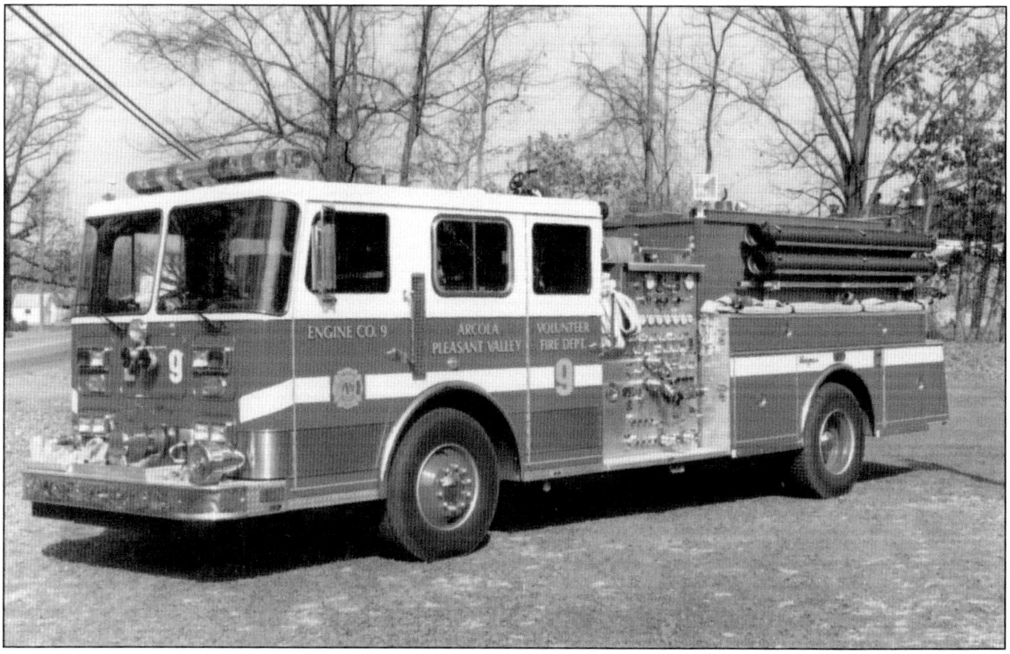

Arcola has had two Seagrave fire trucks as part of its roster. Both the 1990 model and the 1999 Seagrave are four-door engines painted red and white. This four-door 1990 Seagrave came equipped with a 1,500-gallon-per-minute pump and a larger 1,250-gallon water tank. The tank was later replaced with a smaller 1,000-gallon tank.

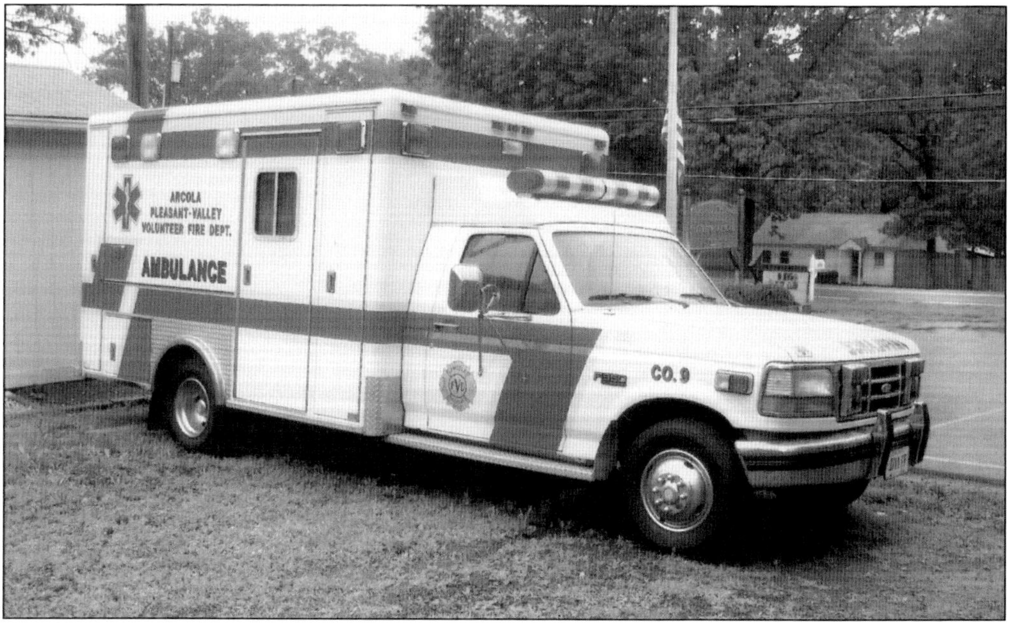

The Arcola Pleasant Valley VFD has owned and operated a variety of Ford box-style ambulances. A Ford/Swab ambulance was placed into service in 1975. The photograph above shows Arcola's 1993 Ford F-350/Horton, which was painted white with red stripes. Arcola would later purchase Ford ambulances built by Horton in 2000, 2003, and 2005 for service at Arcola Station No. 9 and South Riding Station No. 19.

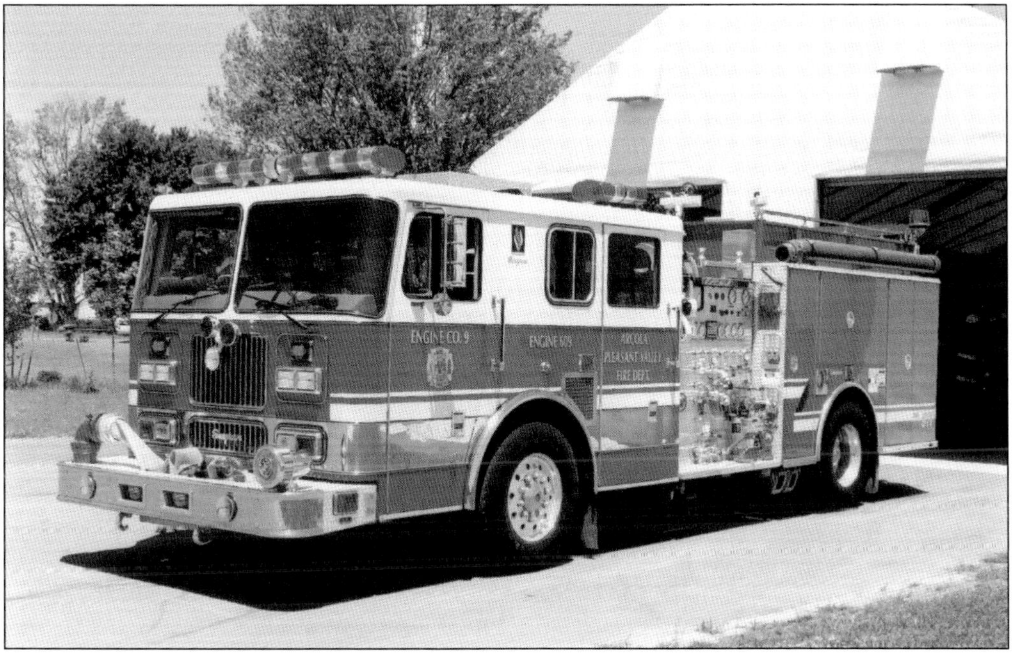

In a joint venture with the Arcola Pleasant Valley VFD and Loudoun County, a new Station No. 19 was opened in South Riding in December 2001. A temporary facility was built but will be replaced with a new public-safety facility in 2007. Apparatus stationed at South Riding include both volunteer- and county-purchased apparatus. Arcola's 1999 Seagrave was photographed in 2003 in front of temporary Station No. 19.

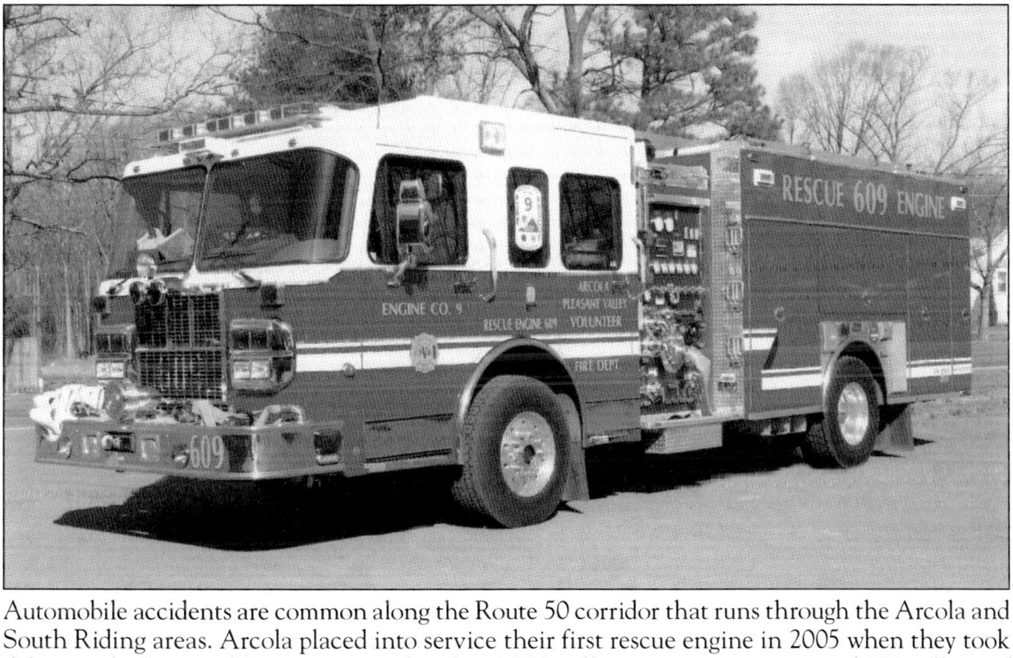

Automobile accidents are common along the Route 50 corridor that runs through the Arcola and South Riding areas. Arcola placed into service their first rescue engine in 2005 when they took delivery of a Spartan/Ferrara equipped with a 1,500-gallon-per-minute pump and 750 gallons of water. In addition to normal firefighting equipment, Rescue Engine No. 9 is also equipped with extrication equipment and other rescue tools.

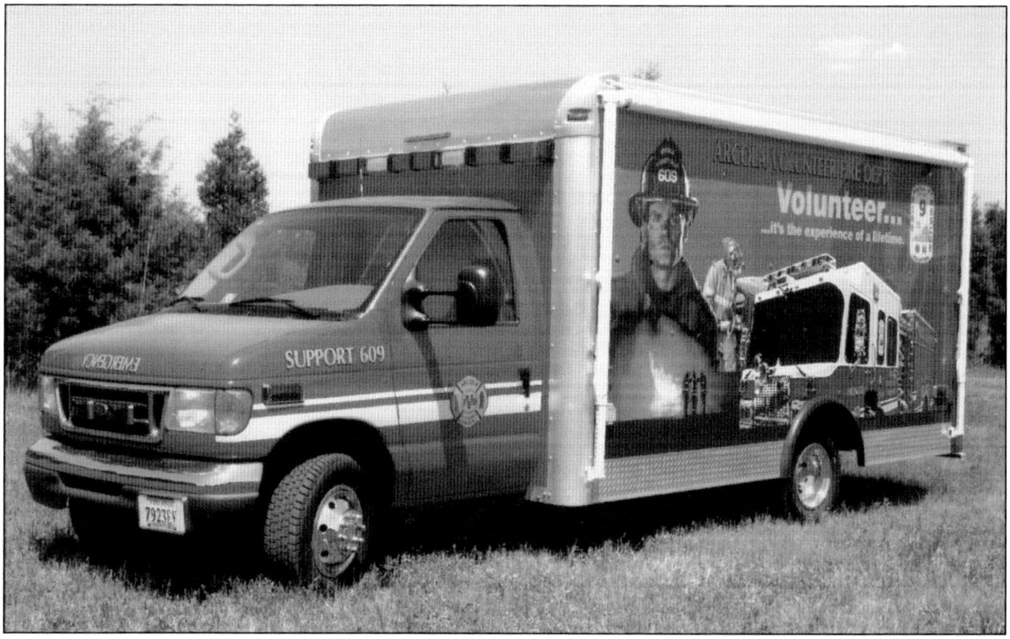

An interesting multipurpose vehicle was delivered to Arcola in 2005 and was designated as Support No. 9. In addition to providing fire and emergency medical support during an incident (canteen, chairs, towels, canopy, portable lights, and such), the 2005 Ford with a Rockport body also acts as a recruitment tool. Recruiting and keeping volunteers is a major initiative in Loudoun County. On the driver's side of Support No. 9 is a graphic design that shows Arcola's rescue engine and has a "fire" look and feel. On the other side of the truck, an Arcola ambulance is on the graphic, and this side has an "EMS" look and feel. The rear cargo door has "Be A Hero—Volunteer" written on it, as well as a list of the benefits of joining the Arcola Pleasant Valley VFD.

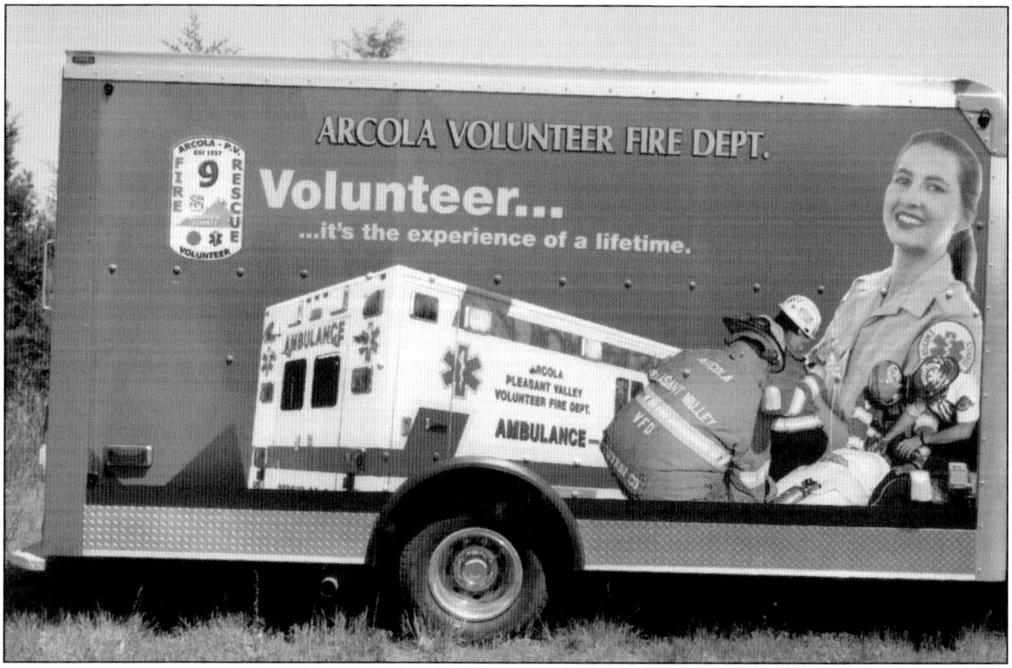

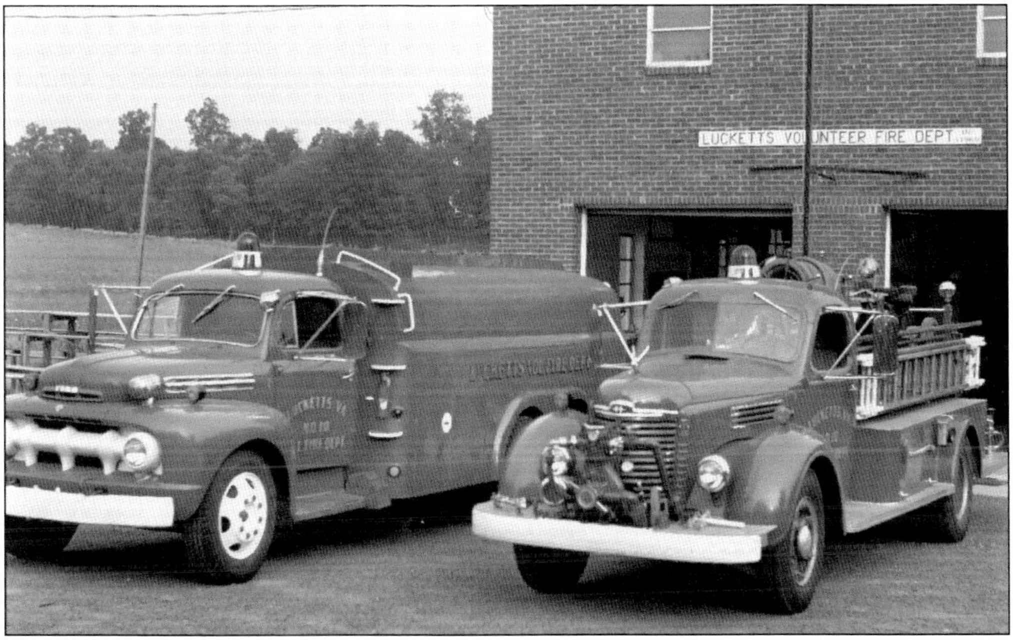

The Lucketts VFD was established in 1960 when the department purchased this 1948 International fire truck (right) for $3,500. Until a new firehouse could be built, the International was kept in a barn. A 1951 Ford 3,000-gallon tanker truck (left) was acquired from Esso and designated as Tanker No. 10. The tanker would later be rebuilt and serve Lucketts for many more years. (Courtesy of Lucketts VFD.)

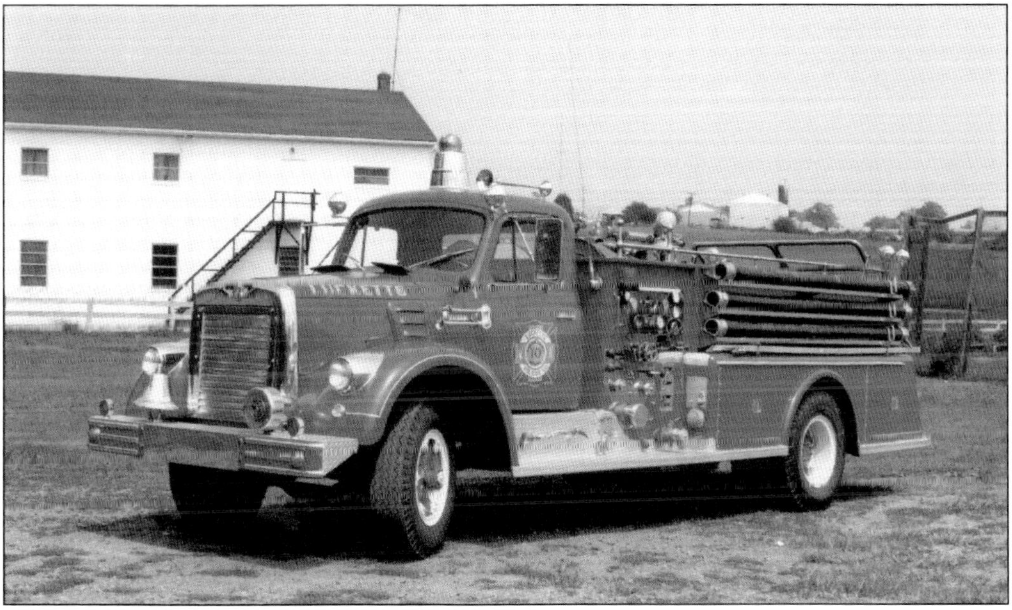

During the mid-1960s, Lucketts purchased this 1959 International/Young 750-gallon-per-minute engine from the Glenn Dale, Maryland, VFD in Prince George's County. Lucketts liked Glenn Dale's department insignia and had a similar one designed with the Lucketts name. To this day, the same Lucketts insignia can be found on a variety of Lucketts's apparatus. (Courtesy of Chuck Madderom.)

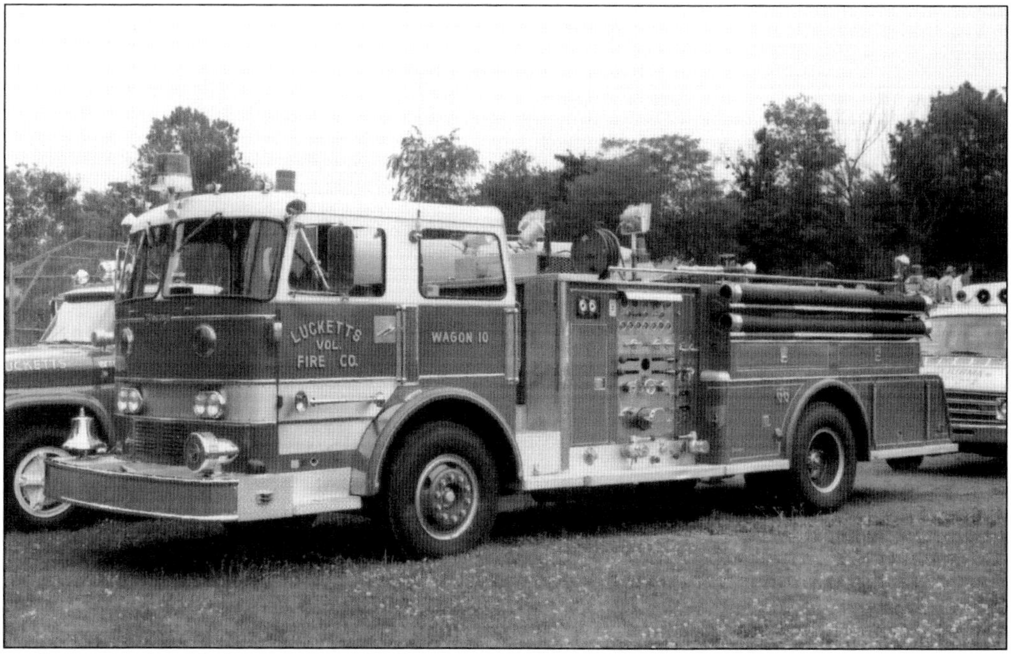

Lucketts purchased this 1970 Duplex/Young engine from the Vienna VFD in Fairfax County, Virginia, and designated it as Wagon No. 10. The Duplex cab was originally an open cab, and Vienna later refitted it with a roof. Wagon No. 10 had a 1,000-gallon-per-minute pump and held 500 gallons of water. Lucketts later sold this engine to a fire department in southwest Virginia.

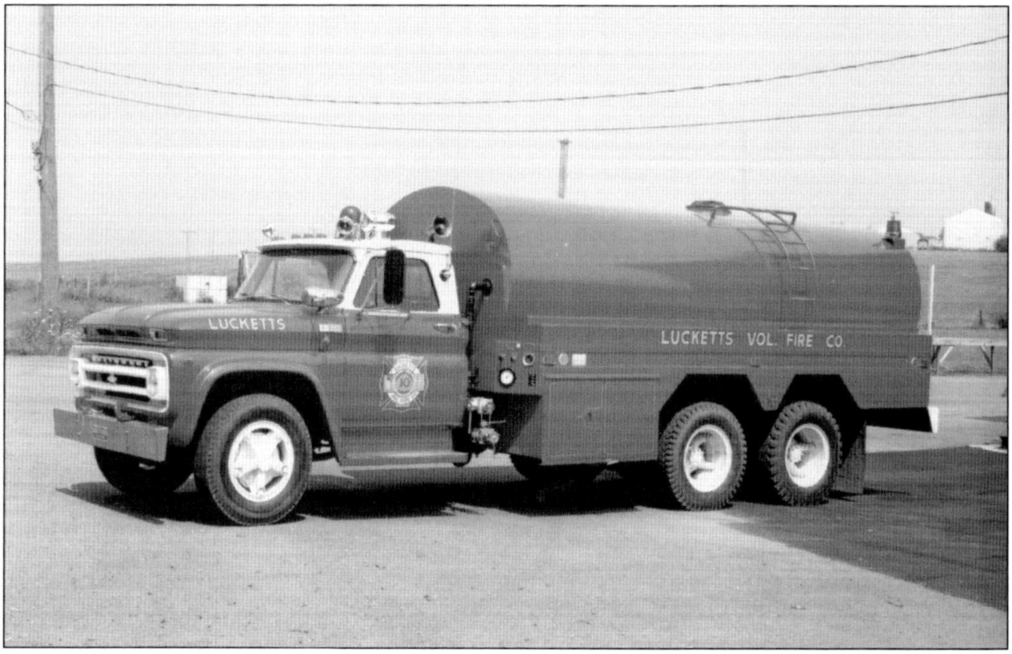

Parts from Lucketts's older 1951 Ford tanker were used to build a replacement tanker in 1977. A 1965 Chevrolet was obtained from a local Lucketts business. 4-Guys built a tanker body onto the 1965 Chevrolet to provide Lucketts with a 3,000-gallon tanker equipped with a small pump. Tanker No. 10 would be rebuilt again in 1990. (Courtesy of Chuck Madderom.)

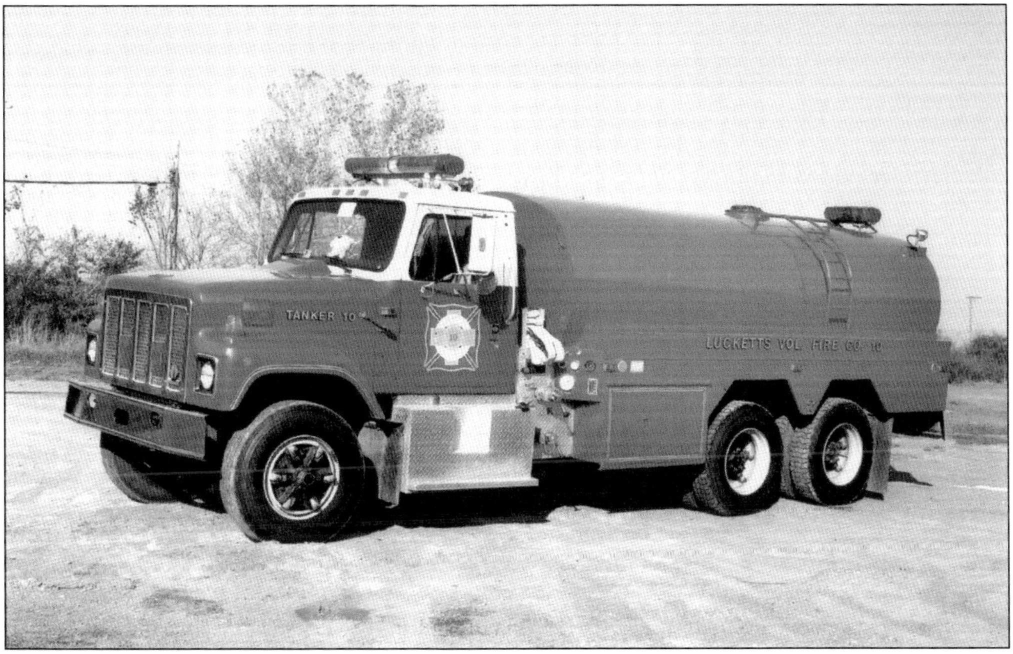

Once again, Lucketts increased the lifespan of their Tanker No. 10 by having the 1965 Chevrolet with a 1977 body by 4-Guys rehabbed. In 1990, an International truck was joined to the older 1977 4-Guys tanker body to provide improved power to the aging tanker. This would not be the last time that Tanker No. 10 would be rebuilt by Lucketts.

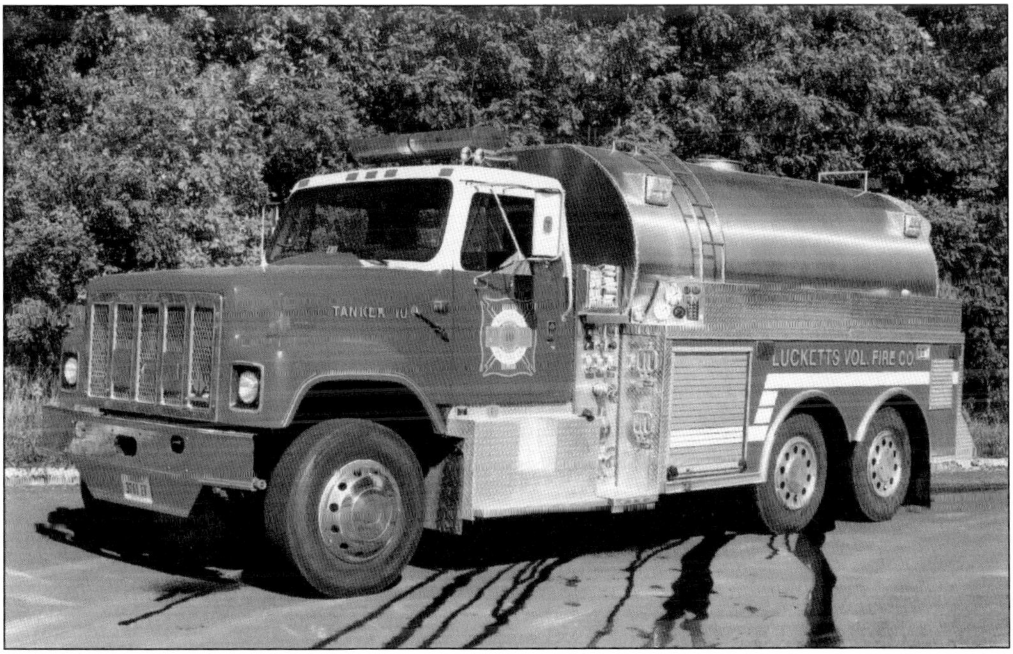

Tanker No. 10 from Lucketts was modified for a third time. Previously the tanker was modified with a new 1990 International cab, but the older tank body was kept. In 1999, the older tank body was removed, and LSI Fire built a new tanker body. Tanker No. 10 is equipped with a 500-gallon-per-minute pump, holds 3,000 gallons of water, and also carries an additional 100 gallons of foam.

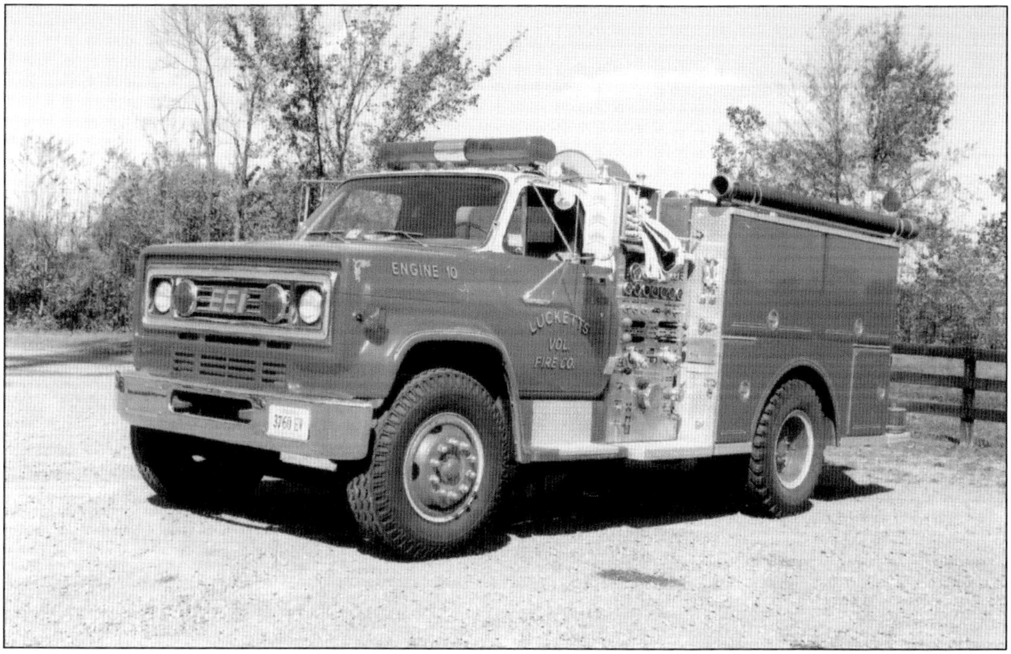

Lucketts purchased this smaller two-man cab engine in 1986. Engine No. 10 is a Chevrolet with a fire-truck body completed by EEI, an old fire-truck manufacturer. Equipped with a 1,000-gallon-per-minute pump and carrying 500 gallons of water, Engine No. 10 was eventually sold to the owner of a farm near Lucketts.

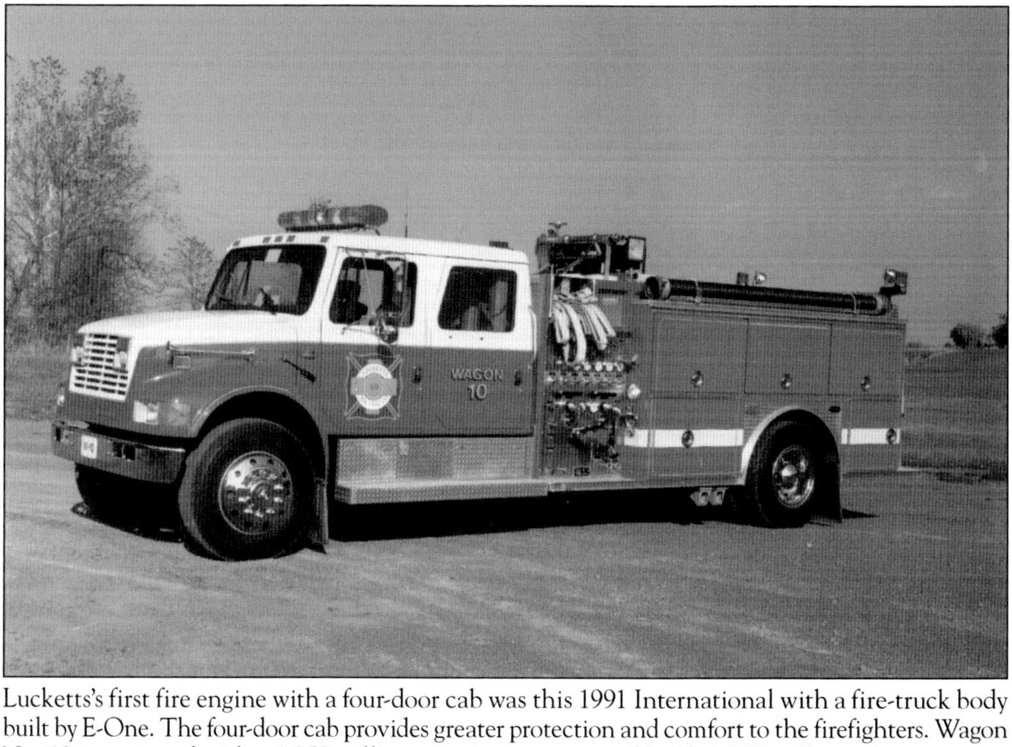

Lucketts's first fire engine with a four-door cab was this 1991 International with a fire-truck body built by E-One. The four-door cab provides greater protection and comfort to the firefighters. Wagon No. 10 is equipped with a 1,250-gallon-per-minute pump and holds 1,000 gallons of water.

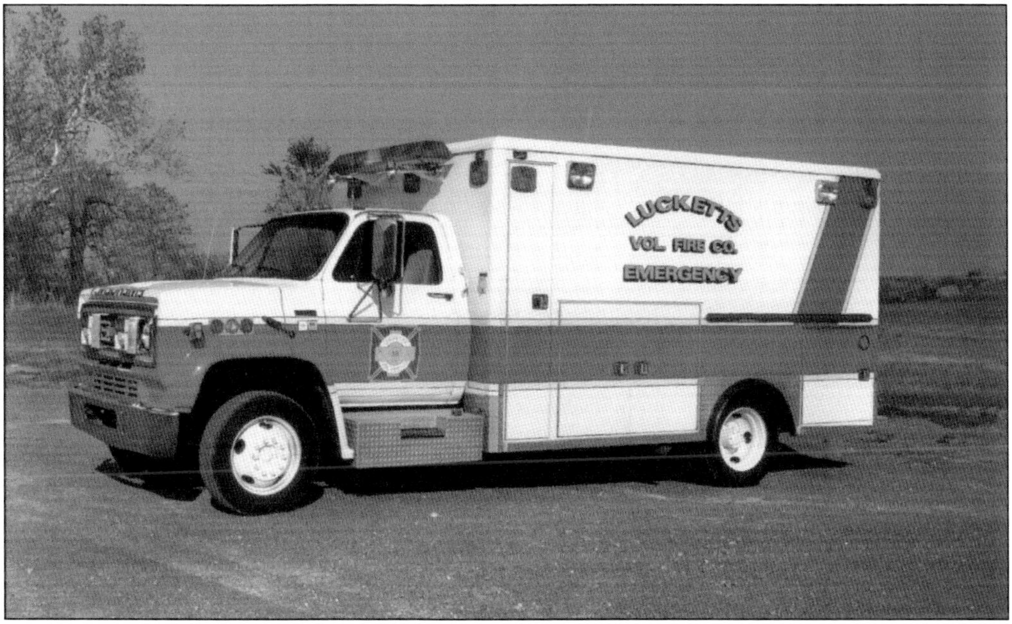

With medical calls rising in the area, Lucketts acquired a variety of old ambulances to be used as first-response vehicles. Sterling Rescue previously owned this 1988 Chevrolet/Wheeled Coach "monster medic," designated Unit No. 10. These vehicles never transported patients to the hospital. They were designed to provide initial care until the arrival of an ambulance. Eventually, in April of 2005, an ambulance was assigned to Lucketts.

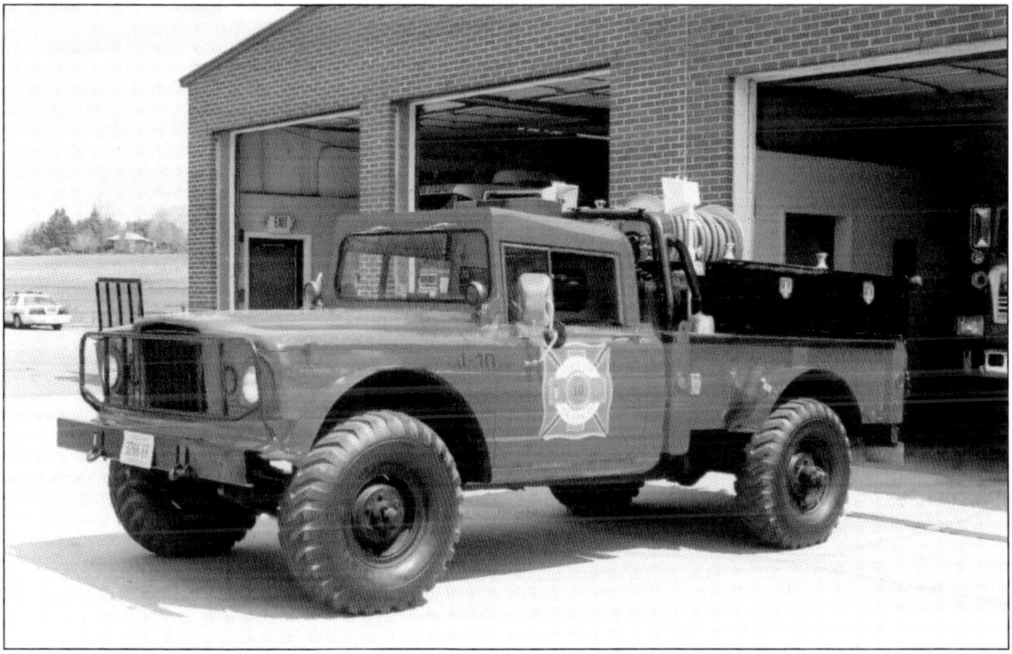

An old military Jeep was rebuilt to become a brush truck for Lucketts. In 1999, the local Greenfield Farms took this 1969 Kaiser Jeep and equipped it for brush firefighting duty in Lucketts. Designated as Jeep No. 10, the Kaiser Jeep is equipped with a 200-gallon-per-minute pump in addition to carrying 200 gallons of water.

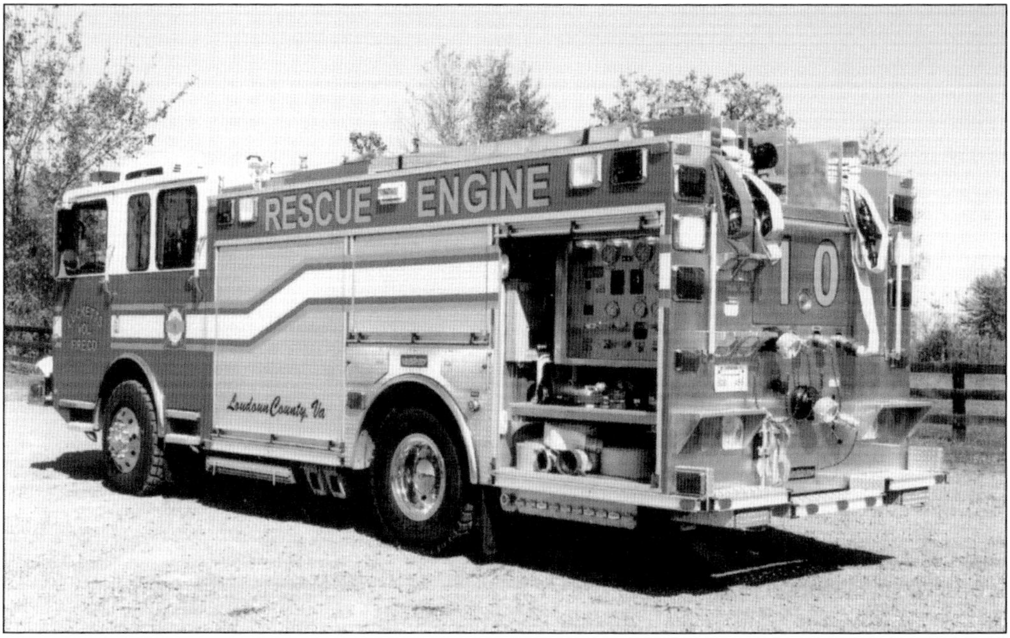

Rescue Engine No. 10 from the Lucketts VFD is quite different from the department's earlier apparatus. The 2002 Spartan cab is equipped with all-wheel drive to help firefighters get to an emergency during inclement weather. Saulsbury built the fire-truck body with the pump panel located in a rear compartment. All hose discharges for the pump panel are located to the rear of Rescue Engine No. 10.

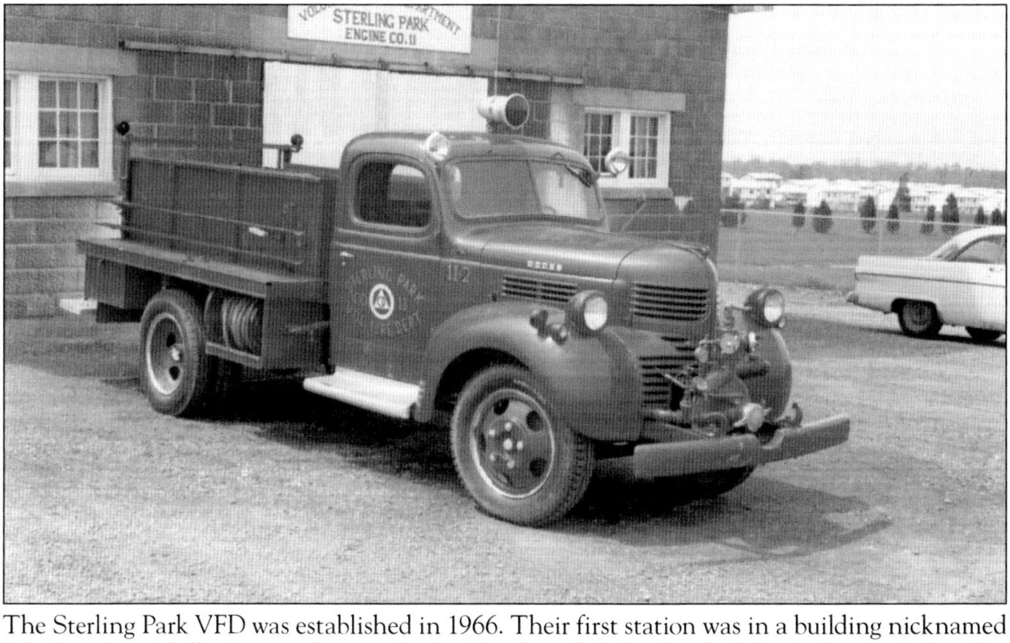

The Sterling Park VFD was established in 1966. Their first station was in a building nicknamed "the barn" on Holly Avenue (currently the Knights of Columbus building). Sterling's first fire trucks were old military and Civil Defense trucks. Photographed in front of the barn is Engine No. 11-2, a 1942 Dodge with a front-mount pump and the Civil Defense emblem on the door. The red Dodge was received from the Round Hill VFD.

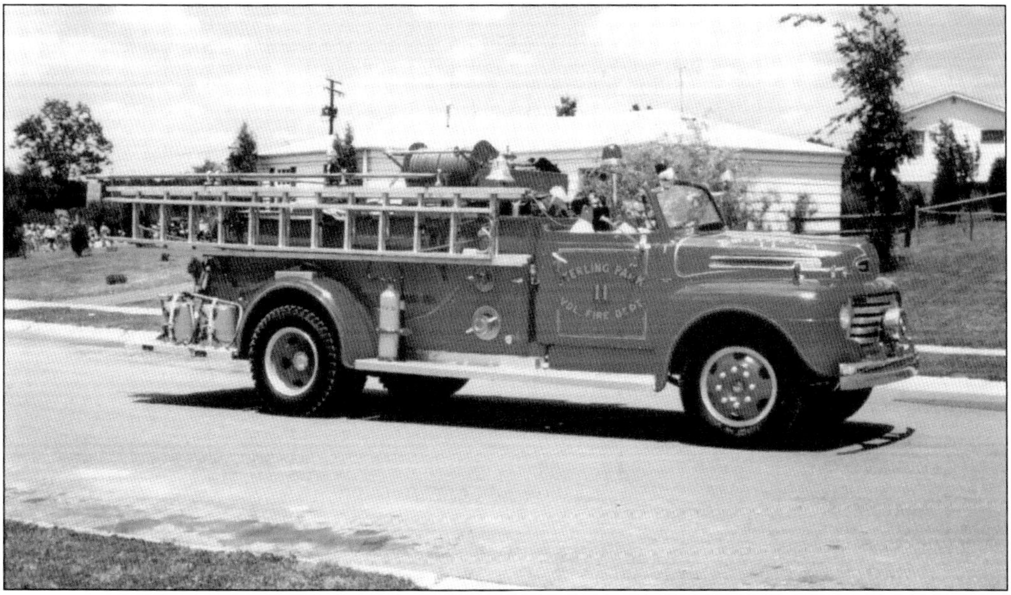

Sterling Park purchased a 1950 open-cab Ford/American LaFrance from the Hillside, Maryland, VFD, located in Prince George's County. The red emergency warning light was mounted on a pole, and a bell was located near the booster reel. After service in Sterling, it was sold to Orlean, Virginia, and it is currently owned and maintained by Dave Hilliard of southern Maryland. (Courtesy of Robert Kimball.)

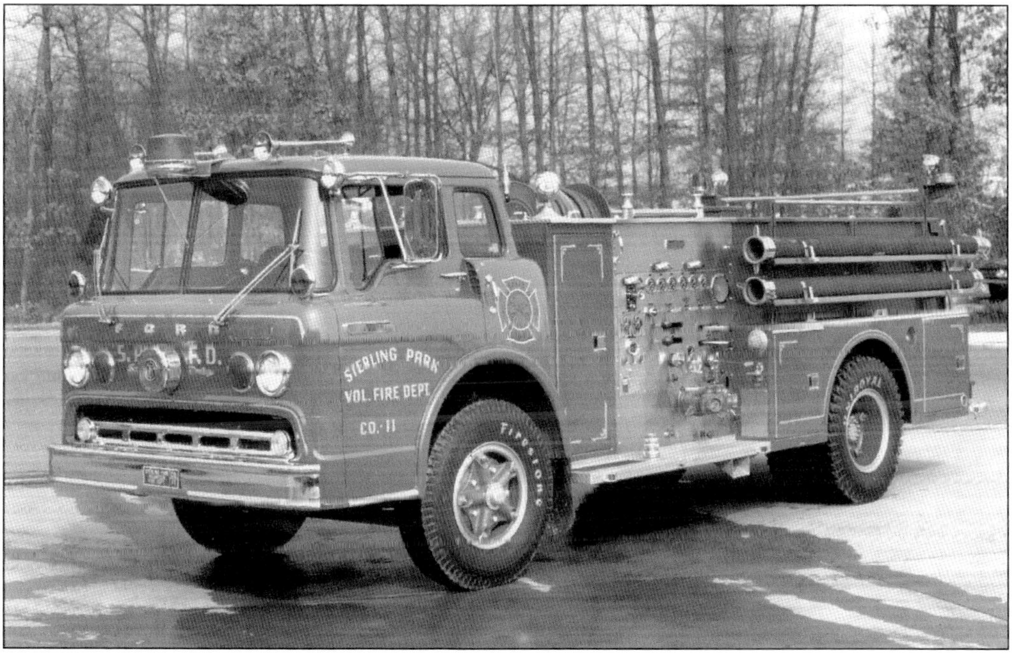

Station No. 11 in Sterling Park purchased two similar 1968 and 1969 Ford C fire trucks built by Ward LaFrance. Originally painted red, they both ran from the barn. The emergency light on the front cab of this 1968 Ford C/Ward LaFrance was mounted lower because of the limited clearance of the barn's doors. Both Fords were later repainted lime green and white. (Courtesy of Robert Kimball.)

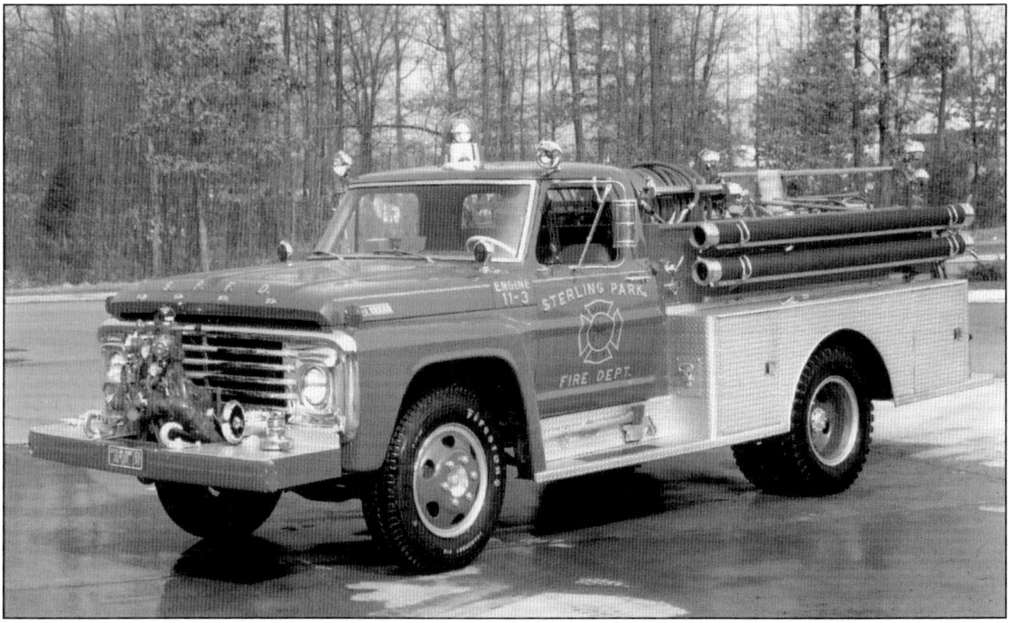

In 1971, the Sterling Park VFD and the Sterling Park Volunteer Rescue Squad moved to the Sterling Safety Center at 104 Commerce Street. Shortly afterward, Sterling Park purchased this 1967 Ford/American 500-gallon-per-minute, front-mount pumper from Marbury, Maryland. Engine No. 11-3 was equipped with 300 gallons of water and later was repainted lime and white. (Courtesy of Robert Kimball.)

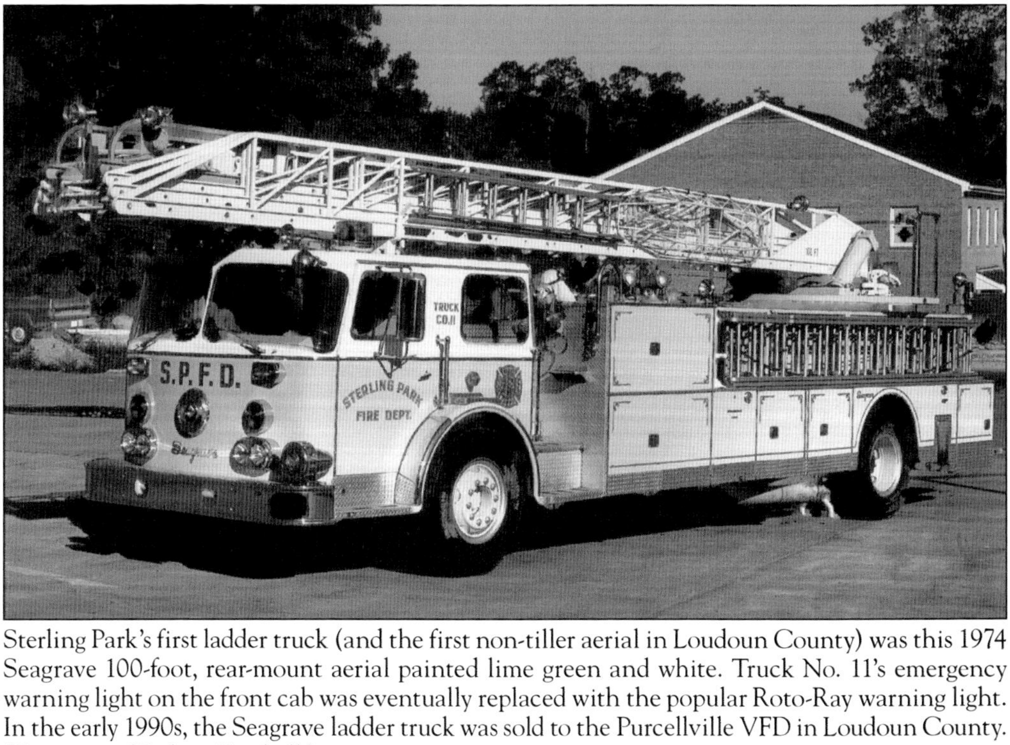

Sterling Park's first ladder truck (and the first non-tiller aerial in Loudoun County) was this 1974 Seagrave 100-foot, rear-mount aerial painted lime green and white. Truck No. 11's emergency warning light on the front cab was eventually replaced with the popular Roto-Ray warning light. In the early 1990s, the Seagrave ladder truck was sold to the Purcellville VFD in Loudoun County. (Courtesy of Robert Kimball.)

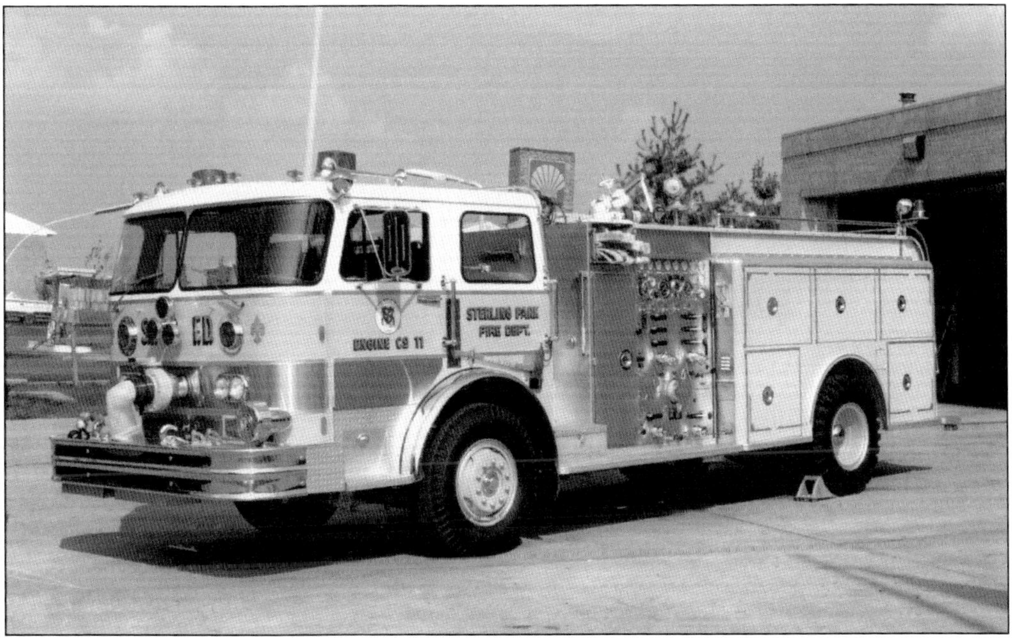

A variety of fire trucks manufactured by Hahn were purchased by Loudoun County fire stations during the late 1970s and early 1980s. Sterling Park's Wagon No. 11 was a 1977 Hahn equipped with a 1,500-gallon-per-minute pump and carried 750 gallons of water. Wagon No. 11 was sold to the Amissville, Virginia, VFD, where it was repainted red and white. (Courtesy of Chuck Madderom.)

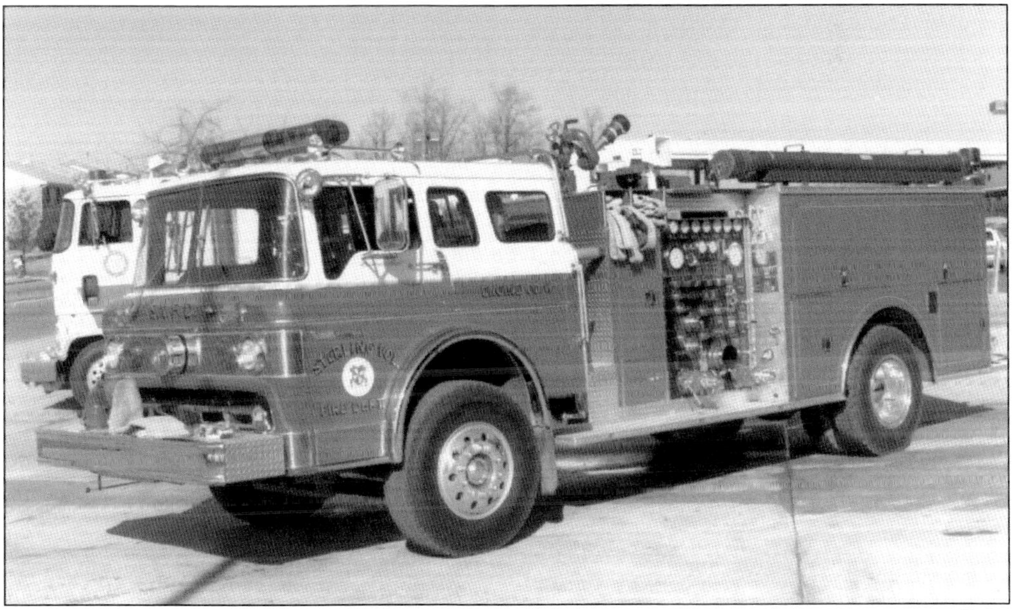

The red-and-white colors returned to Station No. 11 when this 1983 Ford C/Pierce 1,500-gallon-per-minute fire engine was delivered. This Ford was affectionately called the "lead sled," and Engine No. 11 was eventually sold to the Mineral, Virginia, fire department. In addition to the return of red and white, another change to Station No. 11 apparatus was that the trucks were now being labeled as "Sterling" instead of the earlier "Sterling Park."

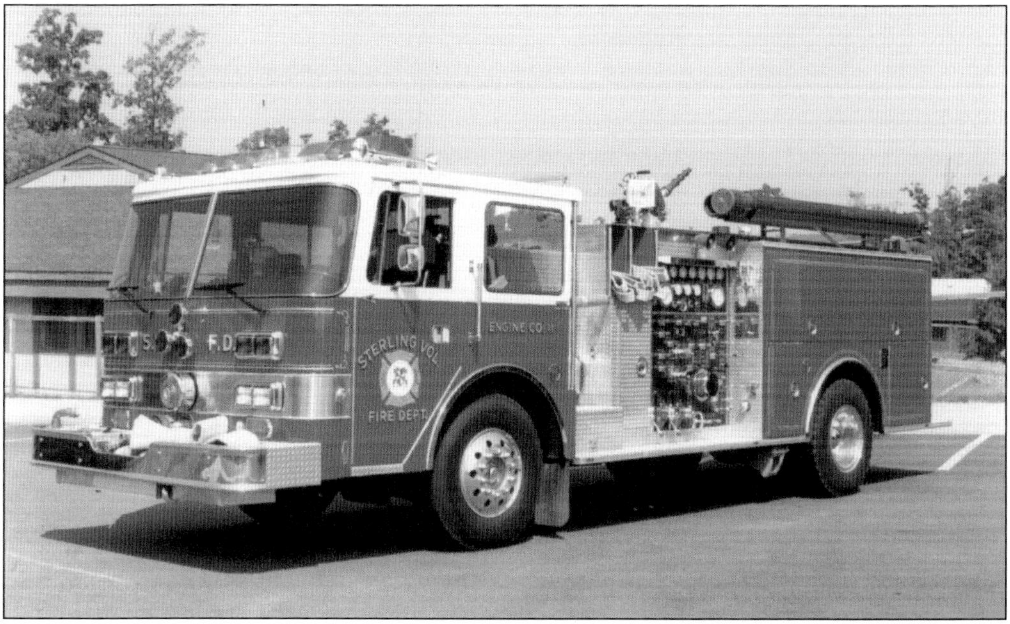

Sterling purchased two similar Pierce Arrow fire trucks in 1984 and 1985. Both engines were equipped with 1,500-gallon-per-minute pumps and held 500 gallons of water. During 1989 and 1992, the Pierces were returned to the Pierce plant in Appleton, Wisconsin, to be rebuilt, which included extending the cab to four doors. Sterling donated both engines to the Loudoun County Fire Rescue Training Center in Leesburg.

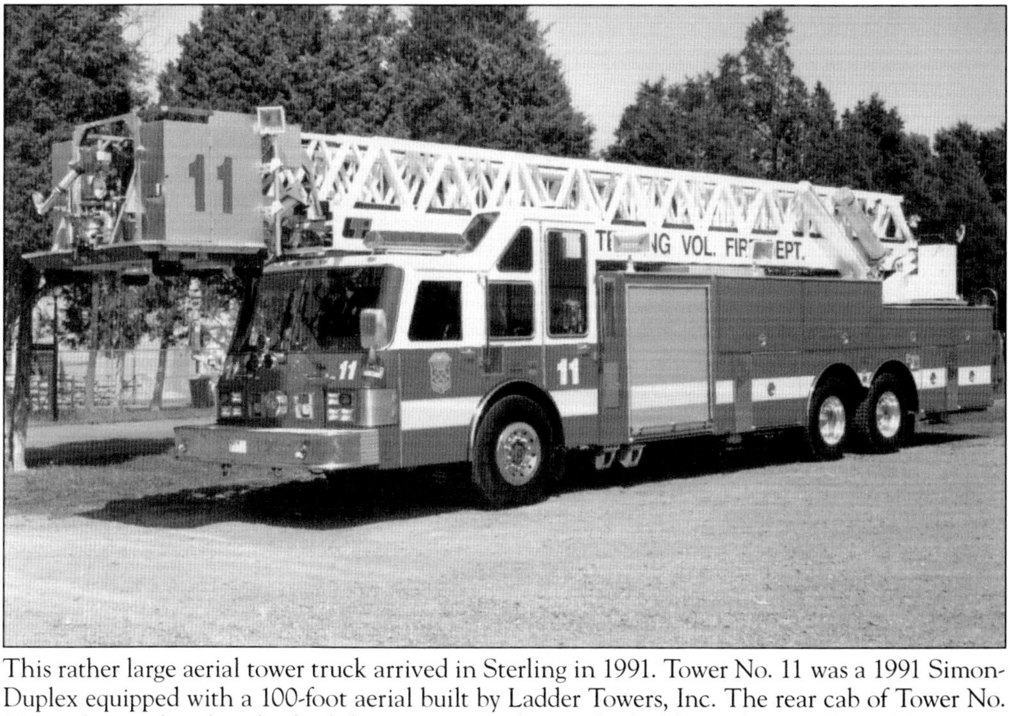

This rather large aerial tower truck arrived in Sterling in 1991. Tower No. 11 was a 1991 Simon-Duplex equipped with a 100-foot aerial built by Ladder Towers, Inc. The rear cab of Tower No. 11 was designed so that the firefighters sitting in the rear had additional room. Tower No. 11 was sold to the Harrisonburg, Virginia, fire department in 2006.

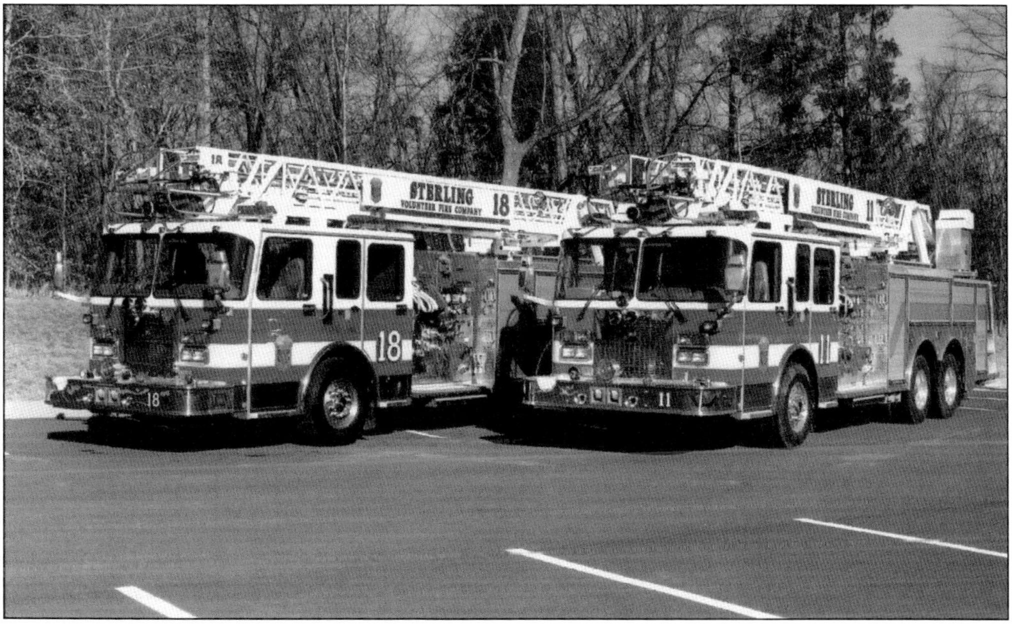

The "Quint" concept was officially tested in Loudoun County when Sterling placed into service Quint Nos. 11 and 18. The Quint could operate as either an engine or a ladder truck, depending on the nature of the emergency. Sterling's Quints were identical 2000 Spartans built by Smeal. They came equipped with 75-foot, rear-mount ladders and 1,500-gallon-per-minute pumps and held 500 gallons of water.

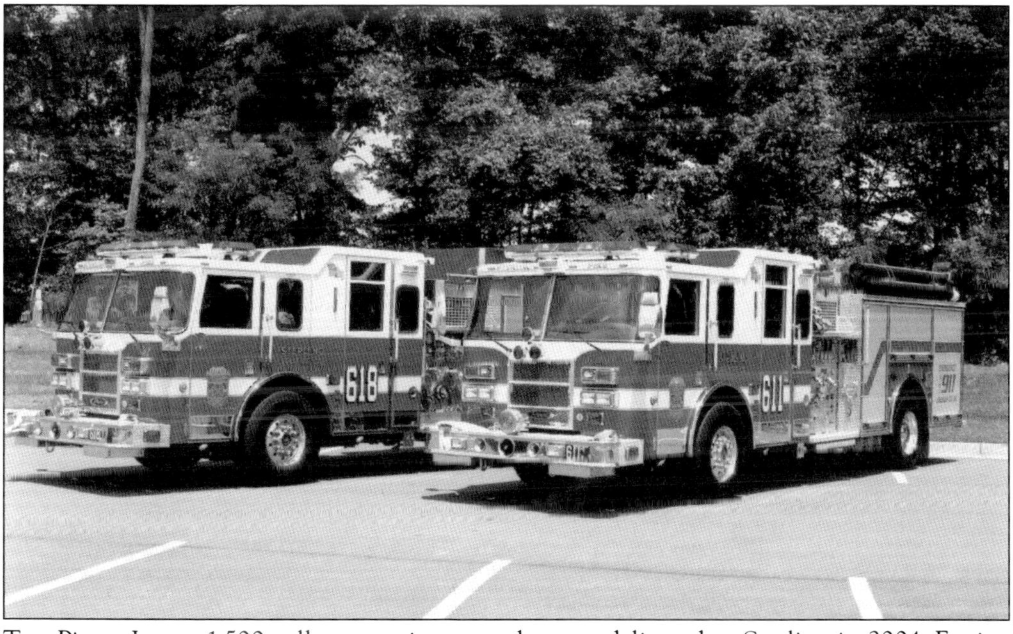

Two Pierce Lance 1,500-gallon-per-minute trucks were delivered to Sterling in 2004. Engine No. 11 is equipped with a 750-gallon water tank, and Engine No. 18 has a 1,250-gallon tank and can operate as a tanker when required. Since Sterling units often respond to Fairfax County, these Pierce trucks also had the No. 6 before their station number, designating them as Loudoun County fire trucks.

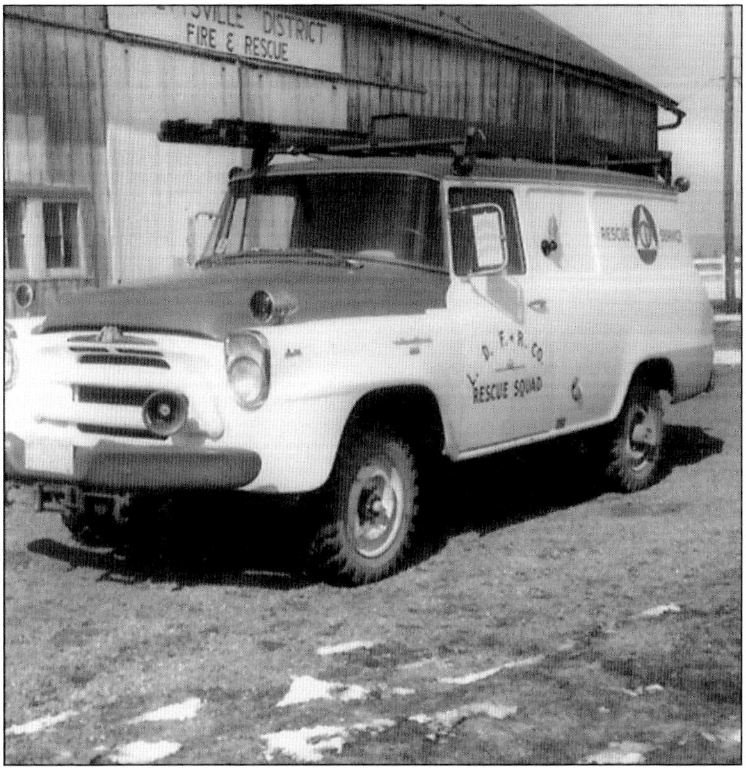

The Lovettsville Rescue Squad was formed in 1966 with a 1955 Ford van loaned to the organization. An early ambulance owned by the rescue squad was a Cadillac ambulance that was purchased for $400 from Graceham, Maryland. The Loudoun County Rescue Squad donated this 1957 International squad truck to Lovettsville in 1967. (Courtesy of Lovettsville Fire and Rescue.)

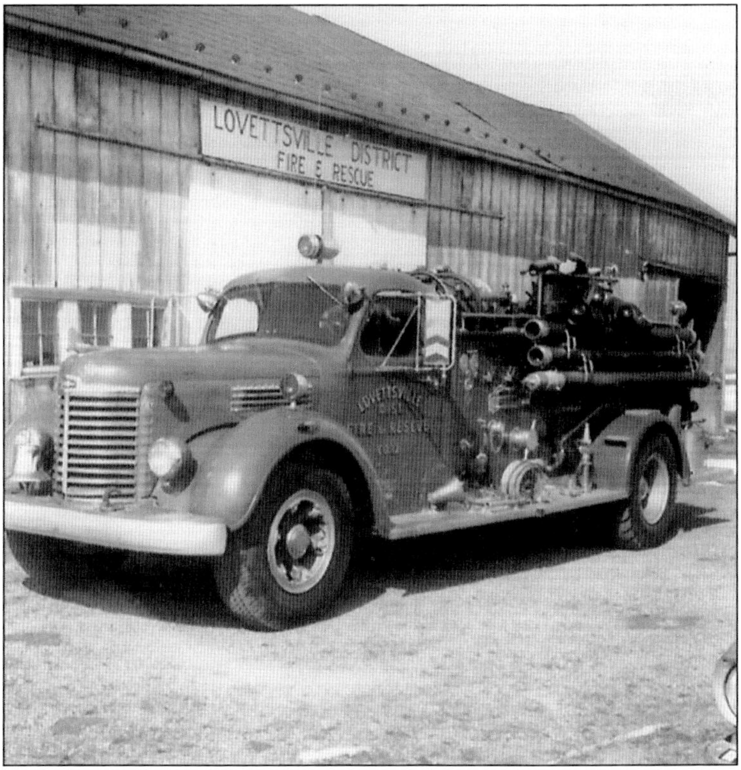

Shortly after the formation of the Lovettsville Rescue Squad, a fire company was started in April 1967. The two departments incorporated as the Lovettsville District Fire and Rescue Company. This 1948 International/American was purchased from Reese, Maryland, and was Lovettsville's first fire truck. It was also Reese's first fire truck. It was photographed in front of a barn that the volunteers initially rented and then purchased in 1968 for $12,500. (Courtesy of Lovettsville Fire and Rescue.)

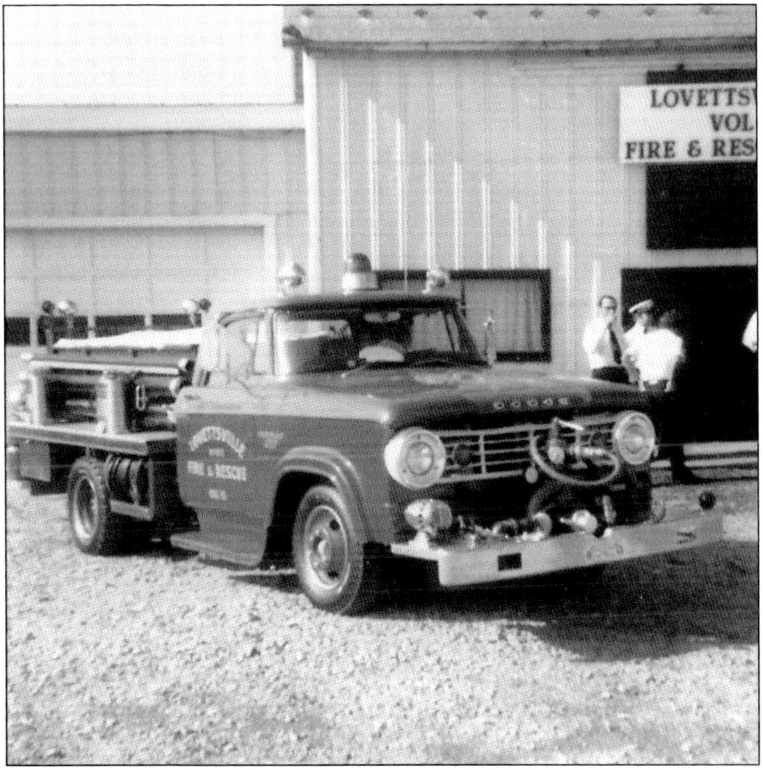

This Dodge from Lovettsville was a rebuild of an earlier Civil Defense fire truck. Originally obtained by Round Hill, it was then used by the new Sterling Park VFD. From Sterling, it went to Lovettsville, where members had the old Dodge cab replaced with this 1968 Dodge cab, which was used as a work truck in a junkyard. In the background is the department's new station. (Courtesy of Lovettsville Fire and Rescue.)

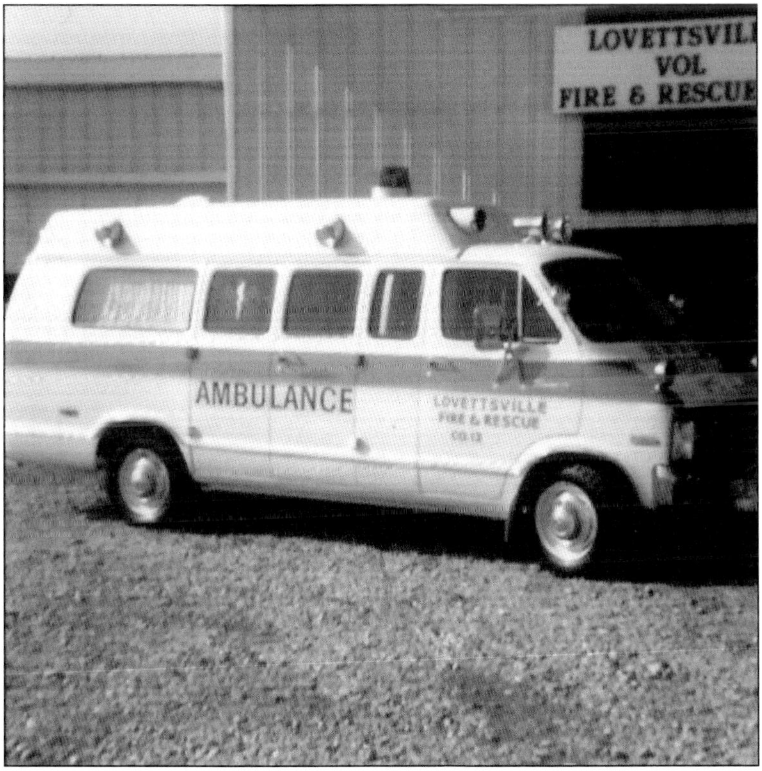

The first new ambulance purchased by Lovettsville was this 1974 Dodge. The purchase price was $13,500. The ambulance was painted white with an orange stripe and was photographed in front of the new station. (Courtesy of Lovettsville Fire and Rescue.)

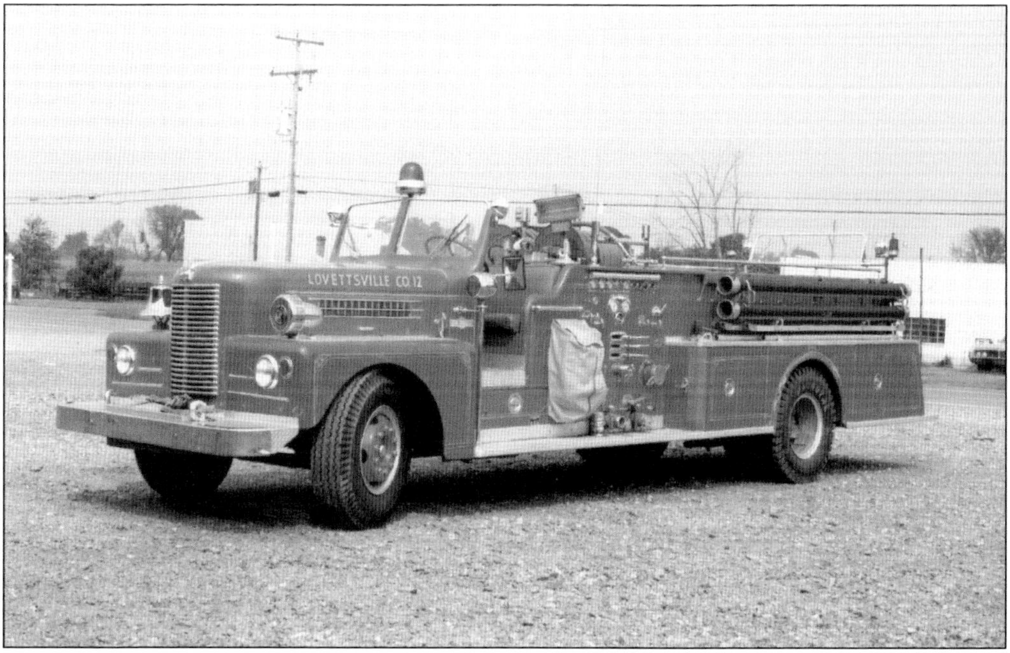

This red open-cab Peter Pirsch fire truck was purchased from the Fairfax City Volunteer Fire Department and ran as Engine No. 12 in Lovettsville. The 1959 Pirsch had a 750-gallon-per-minute pump and carried 400 gallons of water. When its motor blew, the department purchased Hamilton's 1960 front-mount GMC engine to replace the Pirsch. (Courtesy of Chuck Madderom.)

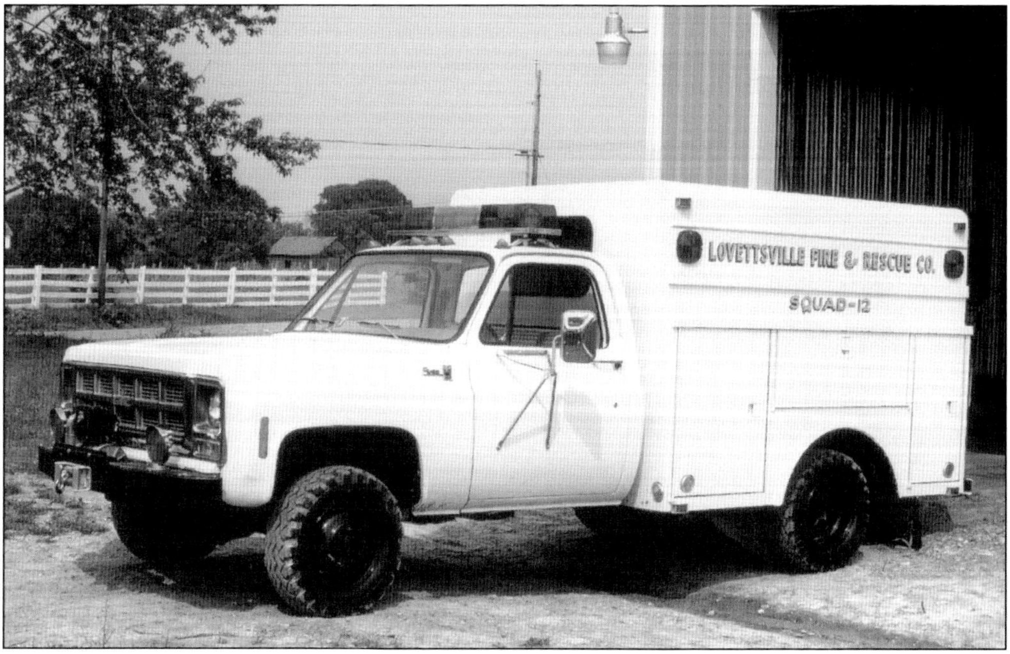

Members of Lovettsville built this 1970s GMC squad truck. The department purchased a new GMC, and members used a Telspa truck body. Painted white, Squad No. 12 replaced the 1957 International squad truck. The squad was photographed out of the rear bay of Lovettsville's current fire station, which opened in 1974. (Courtesy of Robert Kimball.)

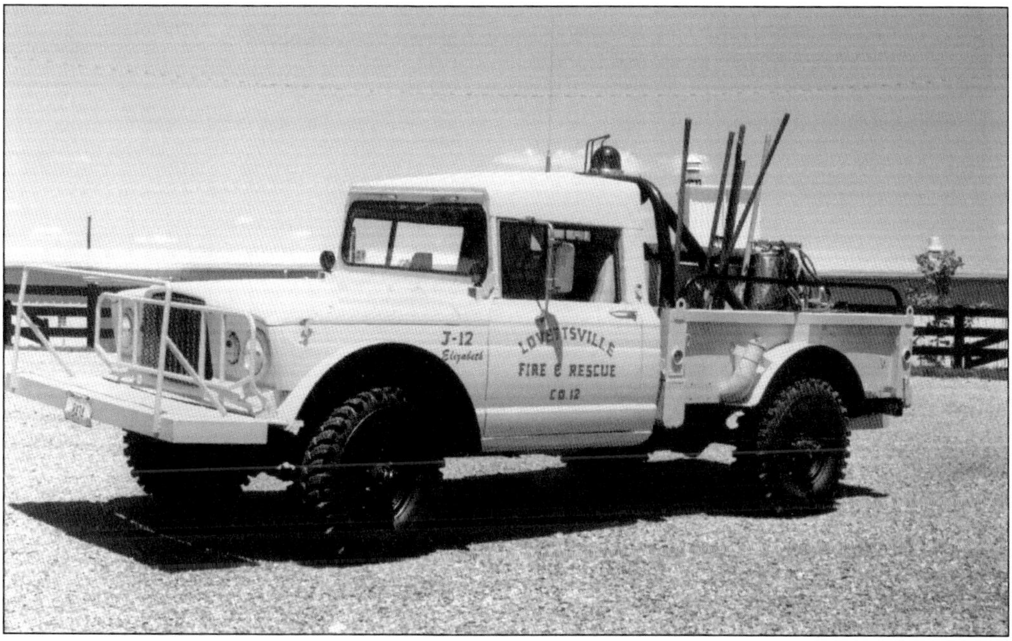

Affectionately known as "Elizabeth," Lovettsville Station No. 12 acquired an old military truck from its neighbors across the state line: the Brunswick, Maryland, VFD. Lovettsville members used the 1968 Kaiser Jeep, built a brush truck, and repainted it to lime green. Jeep No. 12 carries a variety of tools and equipment in addition to 250 gallons of water. It is still owned by the department today.

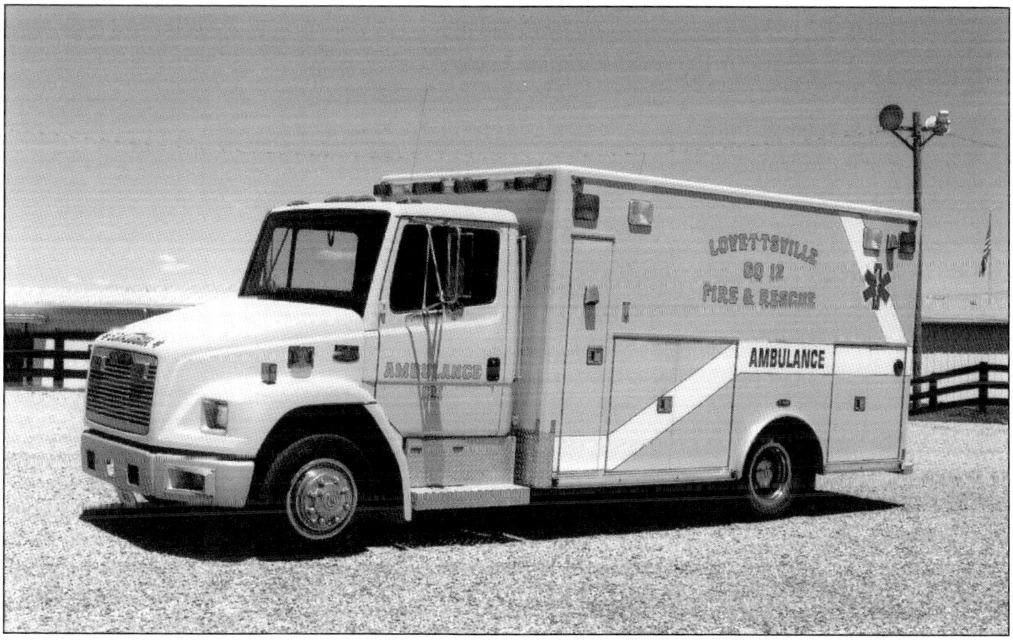

One of Loudoun County's earliest monster medics was Lovettsville's Ambulance No. 12-1, a 1994 Freightliner 60 with an ambulance body built by E-One. Painted lime with a white stripe, the Freightliner was eventually sold in 2002 to the Reva Volunteer Fire Department in Culpepper County, Virginia. Reva repainted the ambulance red and white.

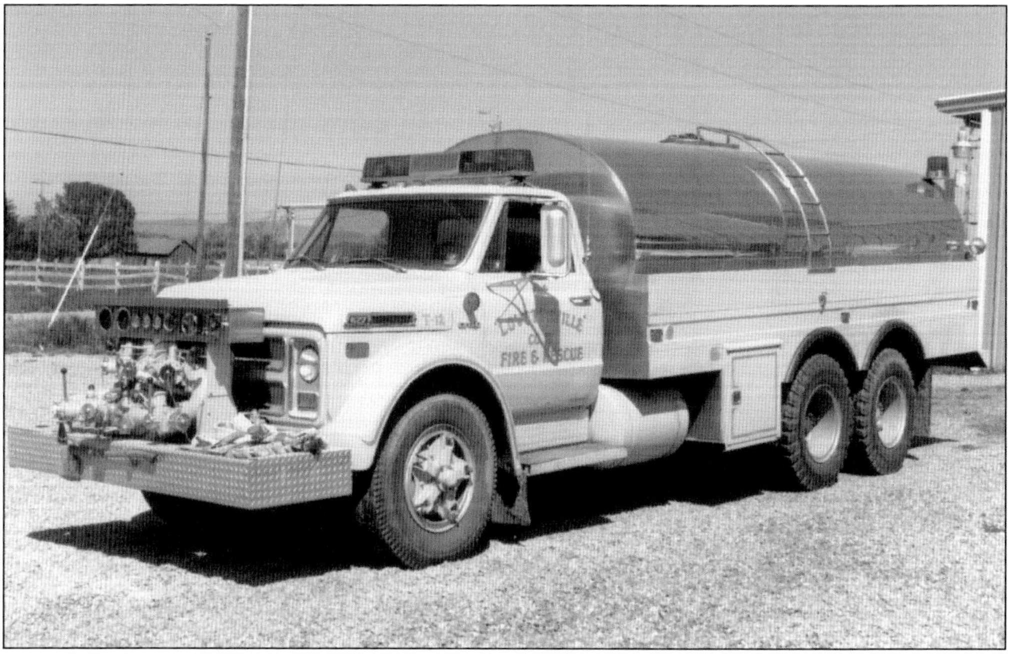

Lovettsville has numerous areas without fire hydrants, so tanker trucks are an important part of the apparatus roster. The department's first tanker was Arcola's first fire truck, the GMC with a 1,250-gallon water tank. In 1978, Station No. 12 purchased a 1972 Chevrolet truck from a coal company. It was sent to 4-Guys in Pennsylvania to be made into a tanker. Tanker No. 12 was equipped with a 1,000-gallon-per-minute, front-mount pump and carried 3,800 gallons of water (above). The Chevrolet was replaced in 1987 when Lovettsville took delivery of their new Tanker No. 12, also built by 4-Guys. The 1987 Mack MC/4-Guys (below) carried 2,500 gallons of water and was sold in 1999 to the Kuttawa, Kentucky, fire department, which repainted the Mack white with blue stripes.

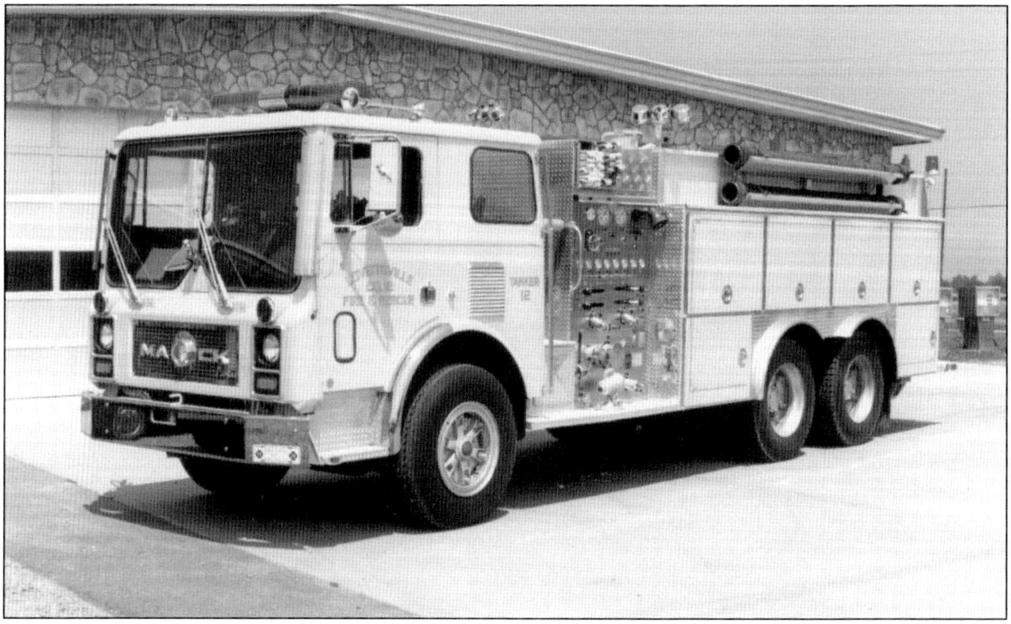

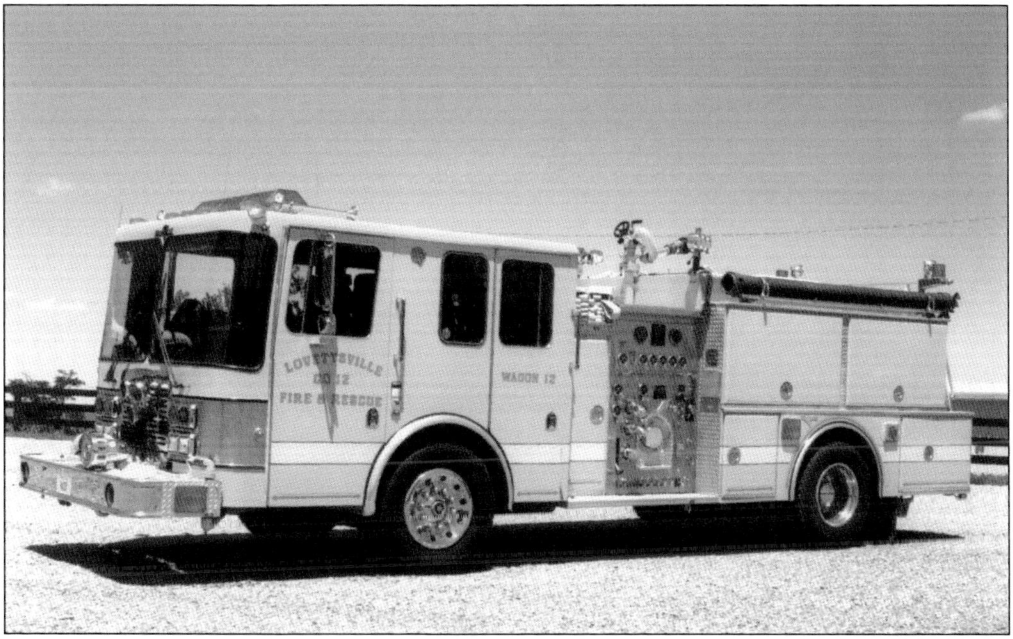

Lovettsville contracted with Grumman, a fire-truck builder in Roanoke, Virginia, to acquire two fire trucks, which were delivered in 1991. Wagon No. 12 was a 1991 Grumman Panther that was equipped with a 1,250-gallon-per-minute pump and that carried 1,000 gallons of water (above). The Grumman was sold to the Six Mile Run Area VFD in Pennsylvania, which repainted the lime-green Wagon No. 12 orange and white. Interestingly the same department also received a 1989 GMC brush truck from Lovettsville and repainted it orange and white. A rescue squad truck built by Grumman was uncommon; however, Grumman did build one at the request of Lovettsville. Squad No. 12, built on an International chassis, also arrived in 1991 and was painted lime green (below).

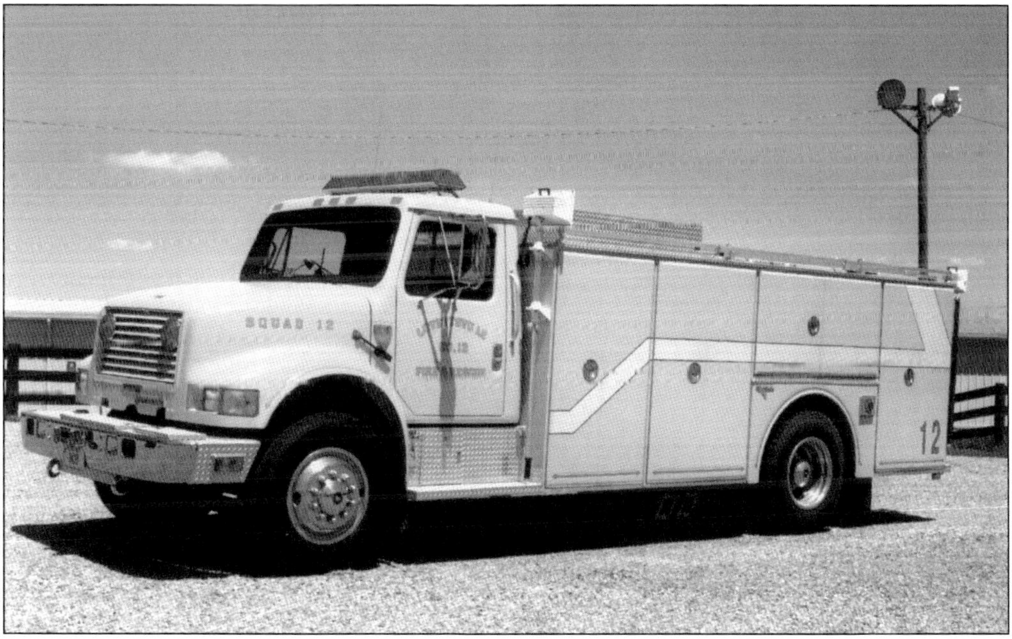

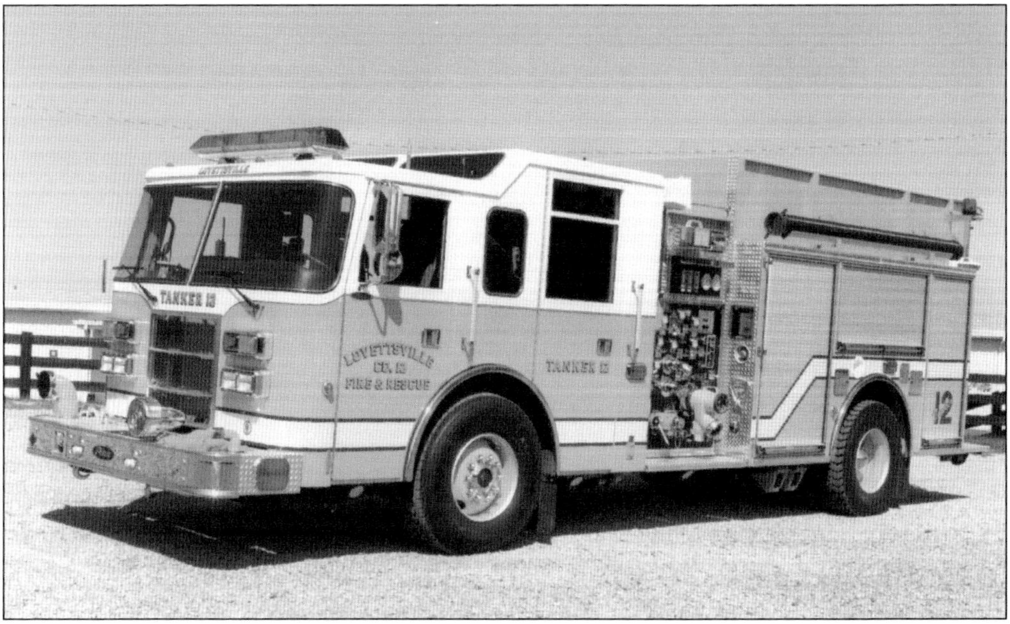

Two Lovettsville Pierce Dash fire trucks were built one after the other at the Pierce plant in Wisconsin. Delivered in 1999, Wagon No. 12 and Tanker No. 12 were painted lime green and white, had roll-up compartment doors, and were equipped with 1,250-gallon-per-minute pumps. Wagon No. 12 was ordered with a 1,000-gallon water tank, and Tanker No. 12 (pictured above) arrived with a 1,500-gallon tank.

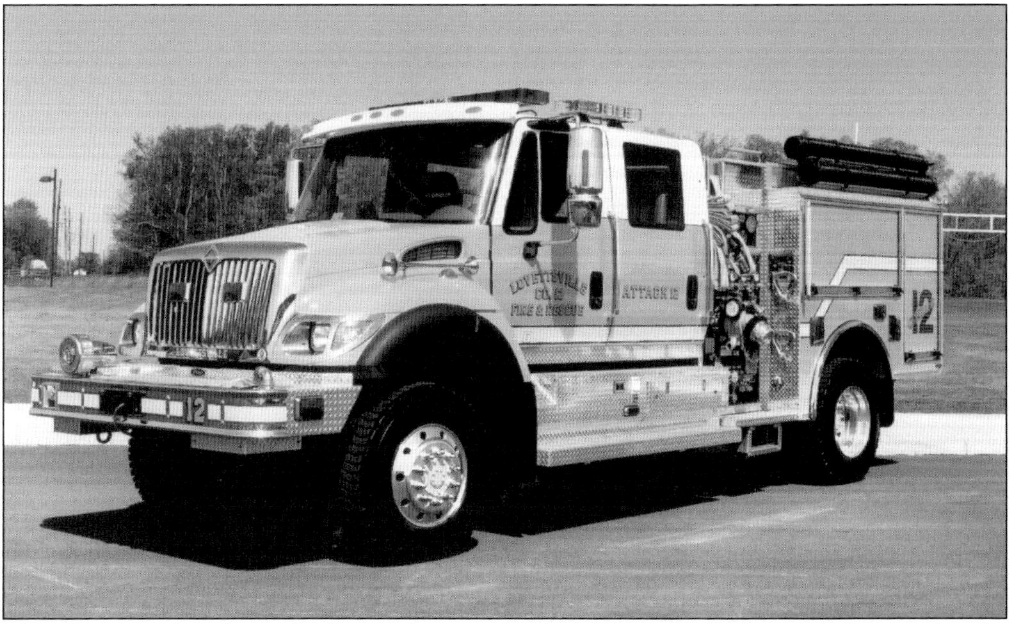

The urban interface design for fire trucks provides the advantage of a fire truck that can handle ordinary fire calls and has the ability to go off-road. Attack No. 12 from Lovettsville is a 2003 International built by Pierce. It is equipped with a 1,000-gallon-per-minute pump and carries 500 gallons of water. Attack No. 12 was photographed at the 2004 Apple Blossom Festival in Winchester, Virginia.

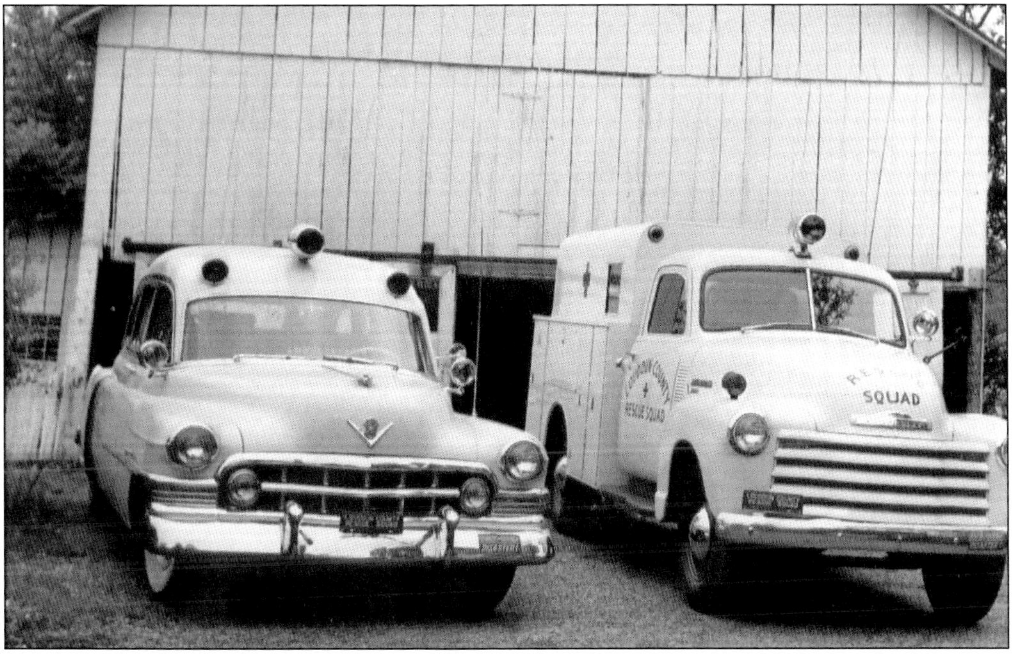

Residents of Leesburg, Hamilton, Lovettsville, and Purcellville organized the Loudoun County Rescue Squad in 1952 with 12 people. Located in Hamilton, the Rescue Squad was given permission from a local grocer to use his 1951 Ford Ranch Wagon as an ambulance. Initially the apparatus sat outside the Hamilton Volunteer Fire Department. Mrs. Rogers, a Hamilton resident, allowed the squad to use an old shed on St. Paul Street as their first station. "The barn" initially had no electricity, and it housed an ambulance and squad truck. The 1955 Cadillac ambulance (left) was previously owned by Loudoun Hospital and was obtained, with the help of Arthur Godfrey, by the squad in 1957. This 1951 Chevrolet (right), known as the "crash truck," was purchased in 1953 and sent to the Cole-Kelly Equipment Company in Richmond to have a utility truck body added. (Courtesy of Loudoun County Volunteer Rescue Squad [LCVRS].)

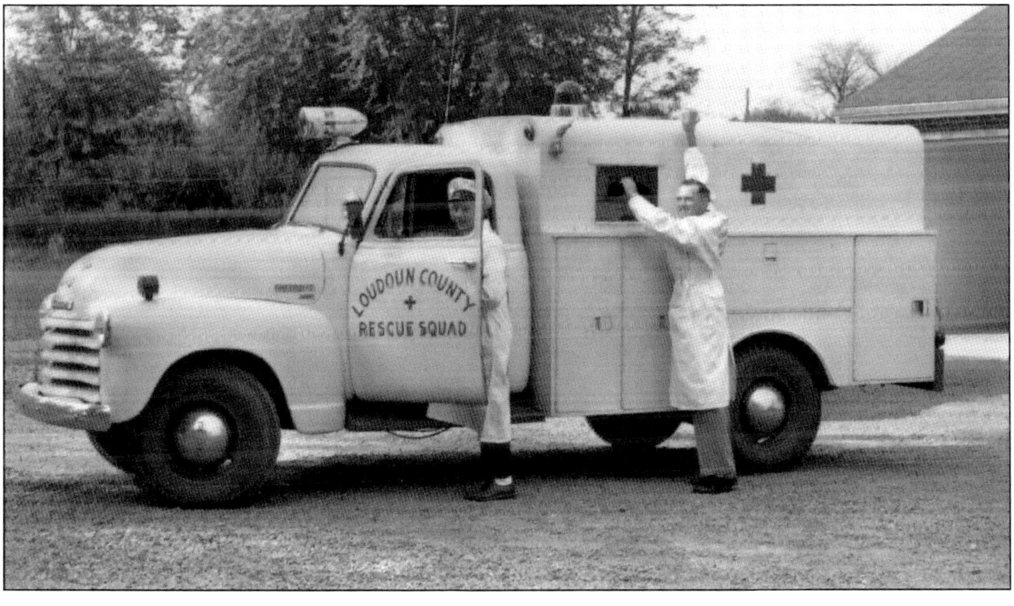

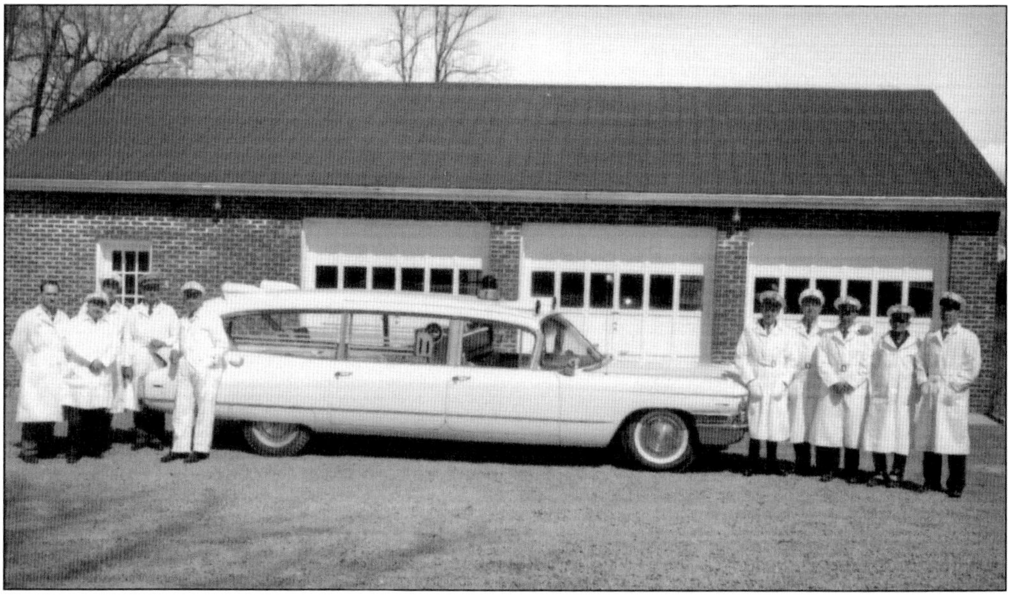

The Loudoun County Rescue Squad built a new three-bay station in 1958 on Laycock Street in Hamilton. A 1960 Cadillac ambulance built by the Superior Coach Company was purchased for $9,500. Local radio and television personality Arthur Godfrey helped obtain the ambulance. Godfrey was a Loudoun County resident and was impressed with the rescue squad after having to use their service after an automobile accident. (Courtesy of LCVRS.)

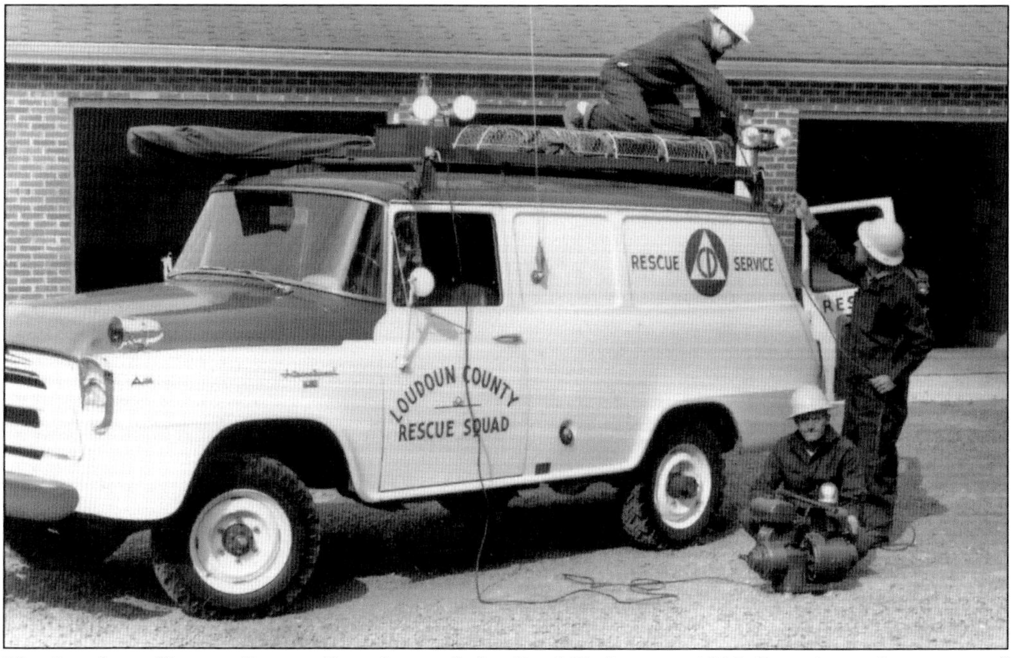

The Loudoun County Rescue Squad joined the Civil Defense in 1954. Squad members trained and prepared for the potential of a nuclear attack during the cold war years. The government donated a 1957 International four-wheel-drive truck to the squad, and it was equipped with a winch, stokes basket, rope, and other equipment. It was later donated to the Lovettsville Rescue Squad. (Courtesy of LCVRS.)

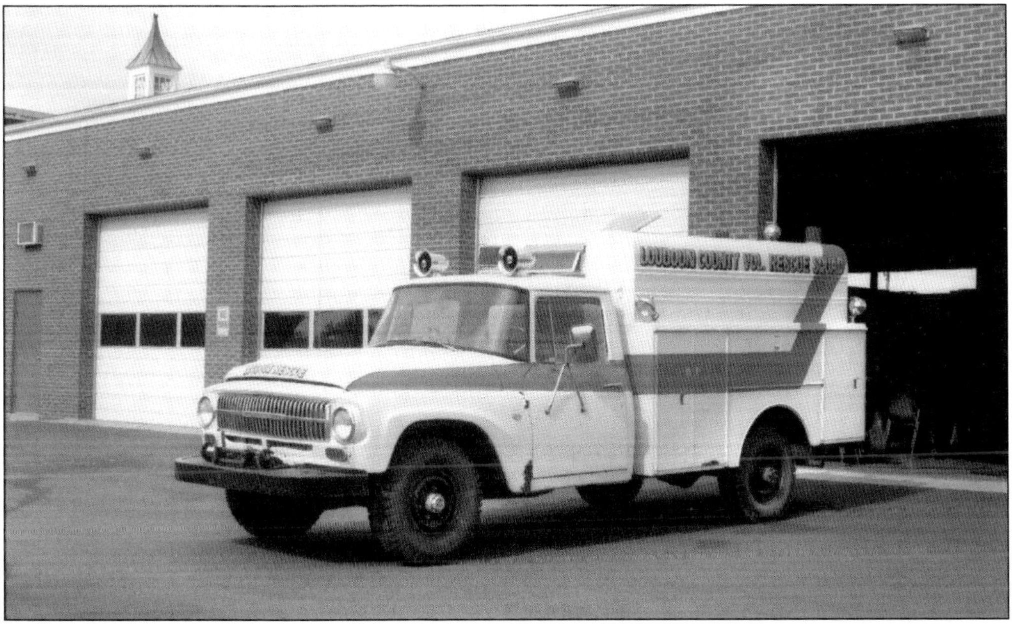

In 1974, on New Year's Day, the Loudoun County Rescue Squad's 1967 International overturned on Route 9 in Paeonian Springs, causing damage to the truck's squad body. The membership purchased a Wilbur body for the International (shown above). The new body was equipped with a small light and intercom, which were used by the crew in the back to tell the driver when the squad could leave the station. (Courtesy of David Bowen.)

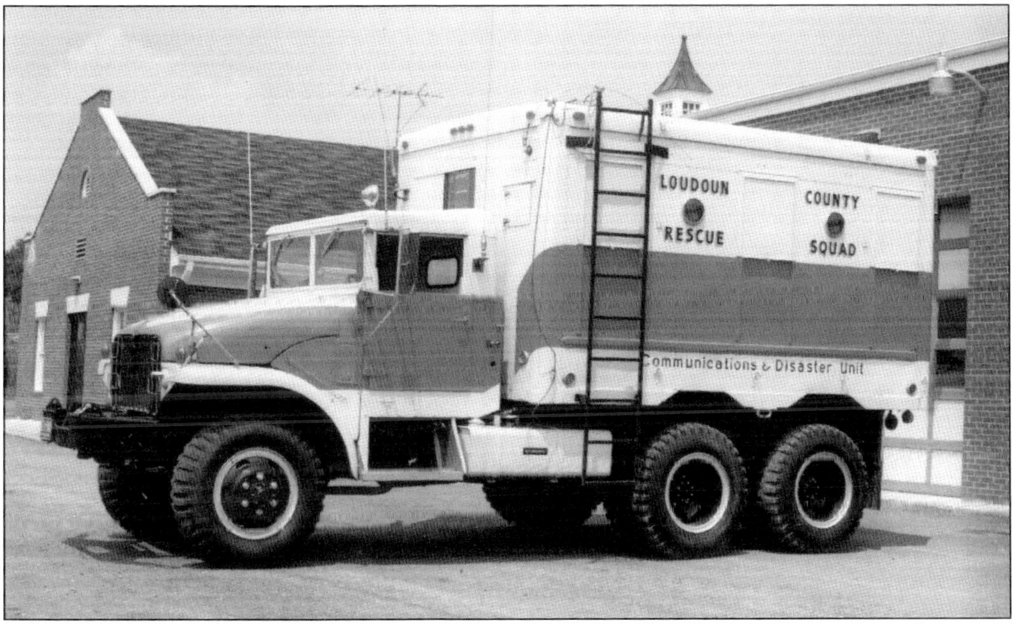

Originally a 1953 military truck, this was the Loudoun County Rescue Squad's communications and disaster unit. A new floor, new cabinets, and other equipment were installed. Before being repainted and while still painted in its silver primer color, it responded to the December 2, 1974, TWA plane crash near Mount Weather, where it was used as a body-removal unit. From that call, it was referred to as the "silver ghost." (Courtesy of Chuck Madderom.)

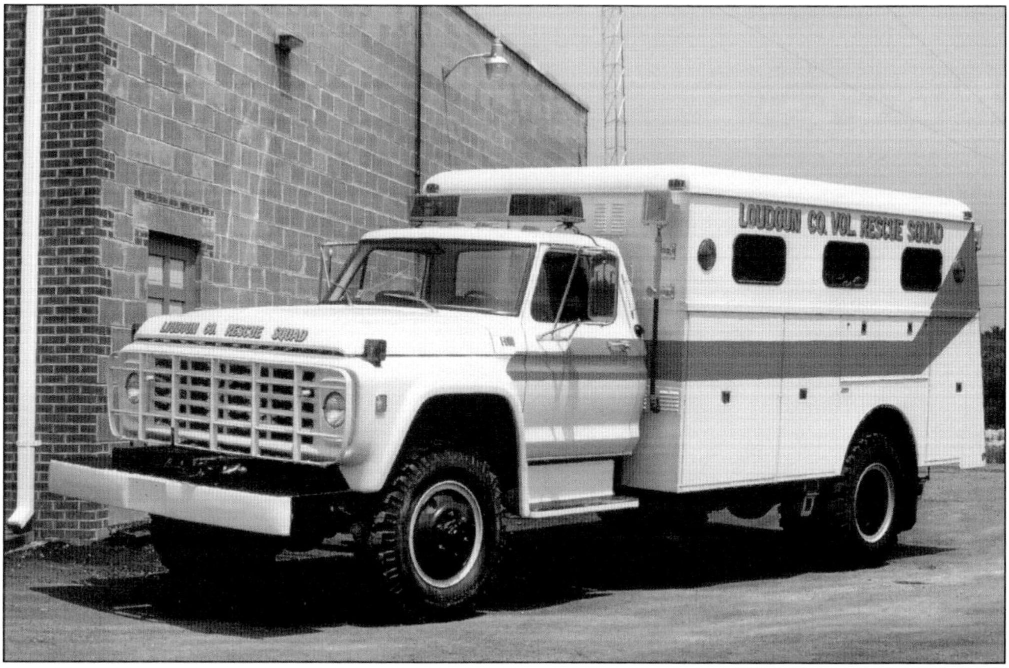

More tools were becoming available to the Loudoun County Rescue Squad, and the squad trucks carrying the equipment became larger. In 1977, the squad placed into service this 1977 Ford F-600 (above), which had a larger squad body built by Reading. The all-wheel-drive capability allowed Squad No. 13 to better access emergency incidents. In the early 1990s, a larger squad replaced the 1977 Ford. Built by E-One of Ocala, Florida, the new Squad No. 13 is a 1992 E-One Cyclone model (below), complete with a four-door raised cab, which gives members of the rescue squad additional room. (Above, courtesy of Robert Kimball.)

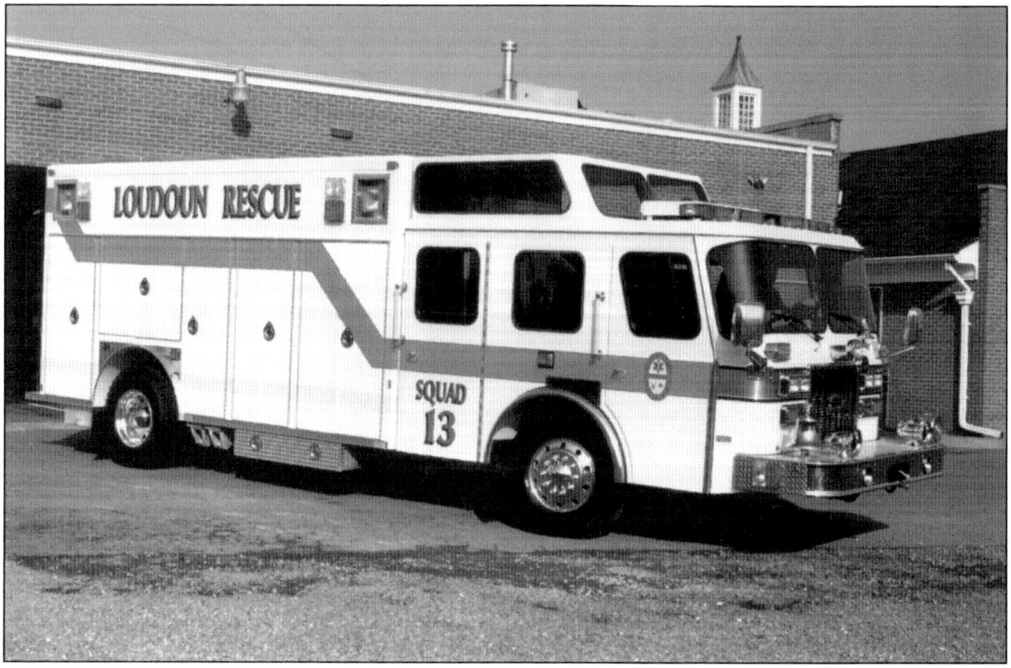

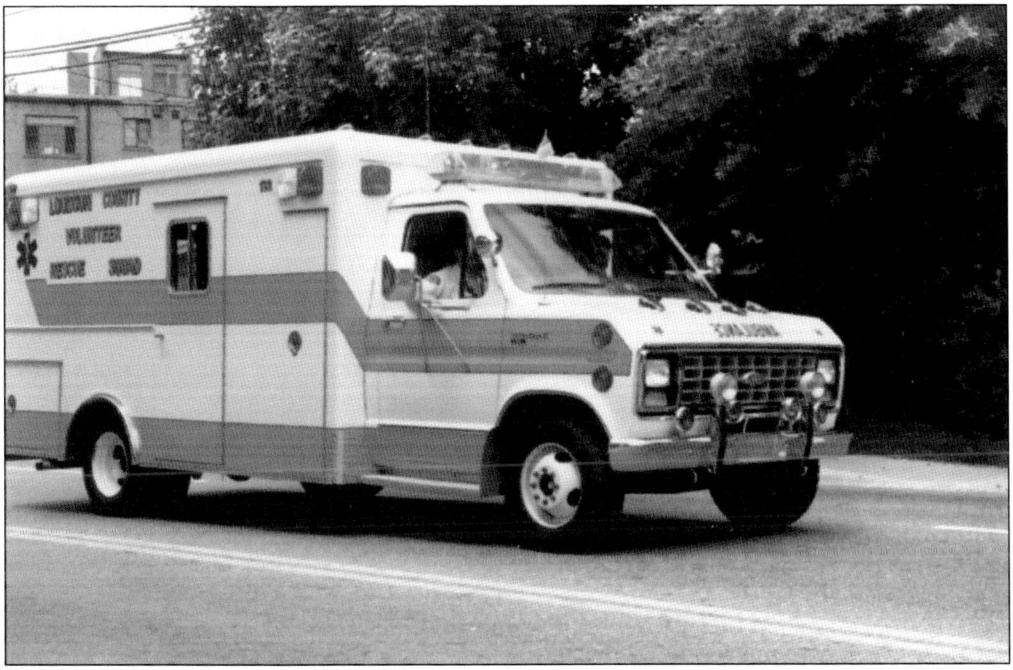

The Loudoun County Rescue Squad used a variety of ambulances during its history, including Cadillacs, Carryalls, vans, box ambulances, and monster medics. Cadillacs were the ambulances of choice from the 1950s through the early 1970s. The squad's last Cadillac ambulance was sold in 1976. Van-style ambulances or a unit with an ambulance box body provided additional room for the crew to work. The squad purchased two box-style ambulances in the early 1980s when they took delivery of identical Fords with ambulance bodies built by Springfield (above). In 1989, a Ford van built by Yankee was added to the Loudoun County Rescue Squad fleet (below). (Above, courtesy of LCVRS.)

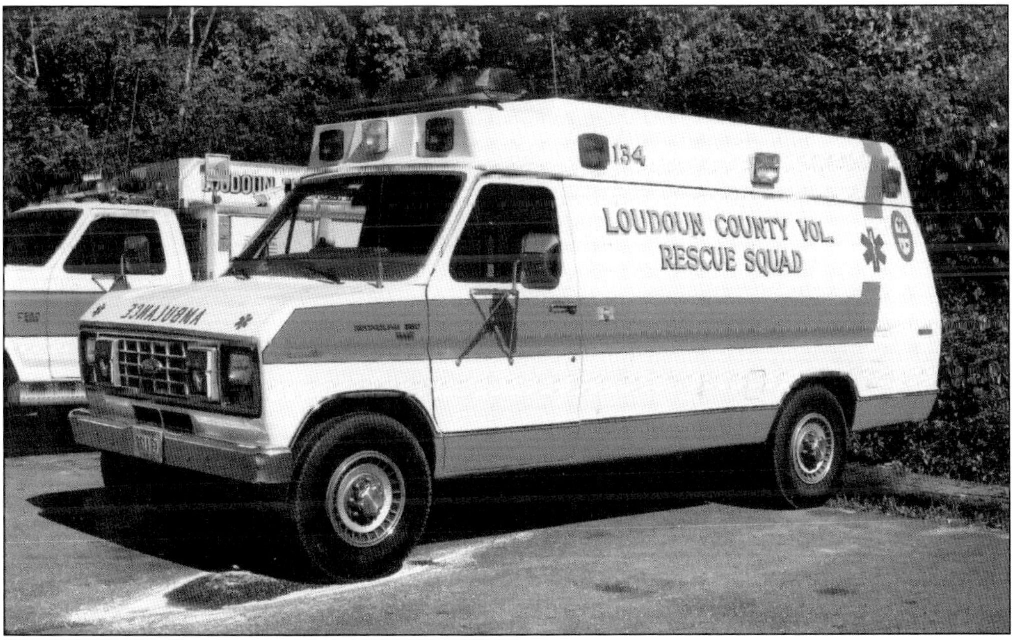

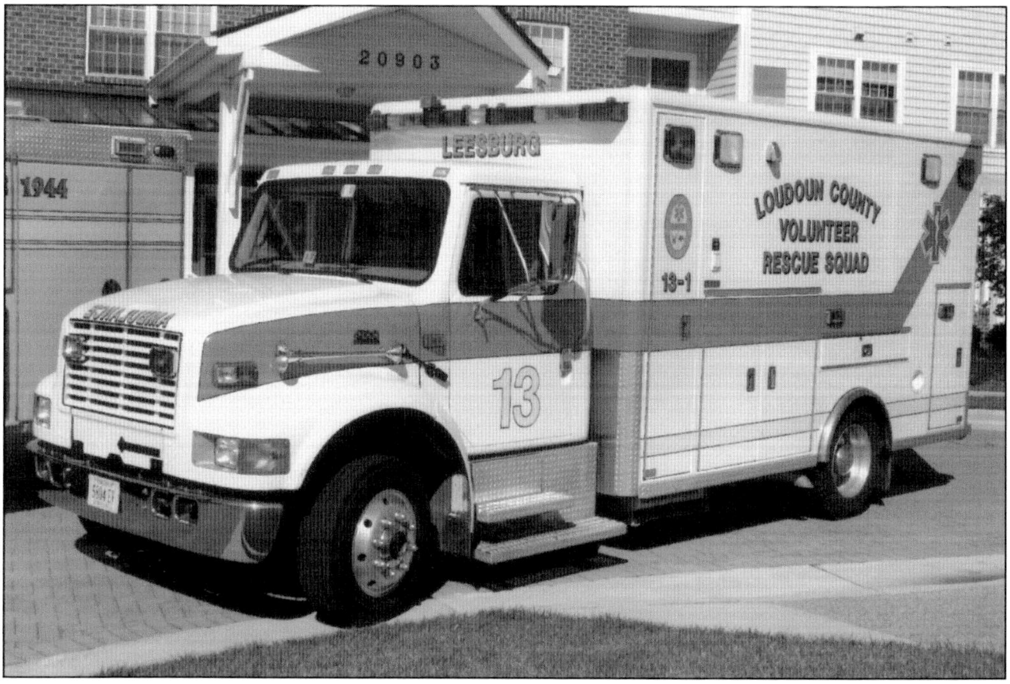

The monster medic ambulance offers many advantages over the traditional ambulance style. Many crews appreciate the additional room in a larger ambulance box and the ability to carry additional equipment. The Loudoun County Rescue Squad has a variety of monster medics on its current roster. Two International 4700 models built by Horton (above) were placed into service in 2000. During 2006, one of the 2000 models was returned to Horton so that a new International 4300 model could be mounted to the older Horton ambulance body. In addition, the squad purchased a brand-new Horton monster medic with an International 4300 cab in 2005 (below).

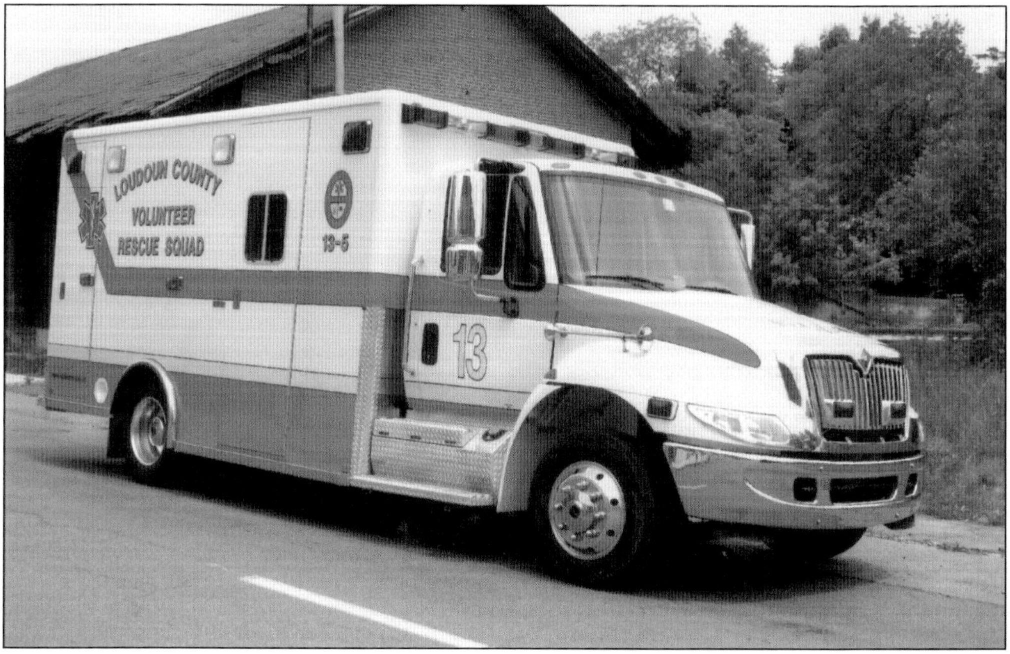

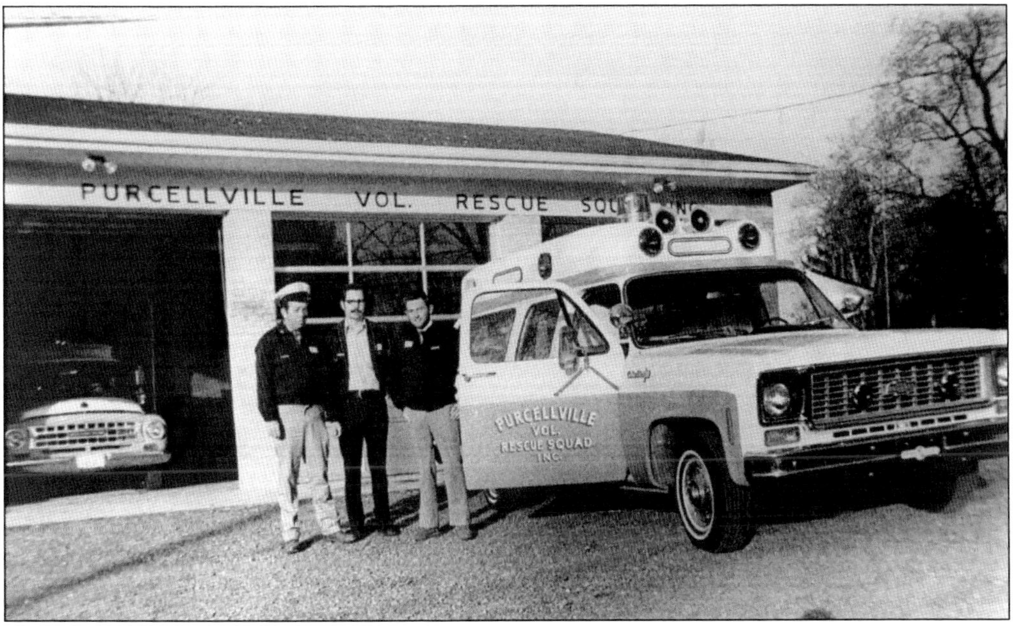

Originally, the Loudoun County Volunteer Rescue Squad had an ambulance available in Purcellville. A new independent Purcellville Volunteer Rescue Squad began operations in 1969 by using an old mechanics shop at 211 South Twentieth Street. This Chevrolet Carryall ambulance was previously a Loudoun County Rescue Squad ambulance. The nose of a 1969 International ambulance can also be seen. (Courtesy of Purcellville VRS.)

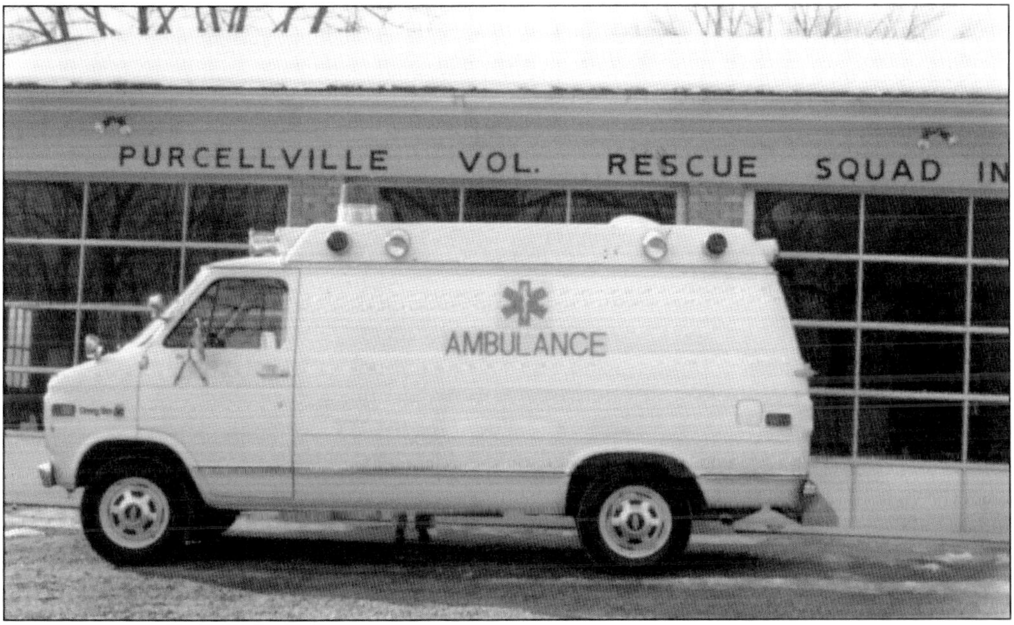

The Purcellville Volunteer Rescue Squad purchased this new Chevrolet van-style ambulance in 1977. It was ordered from an ambulance manufacturer in Florida. It arrived in Purcellville painted white. It eventually had a red stripe and Purcellville graphics added. This photograph shows the ambulance in front of the rescue-squad station shortly after arriving in Purcellville. (Courtesy of Purcellville VRS.)

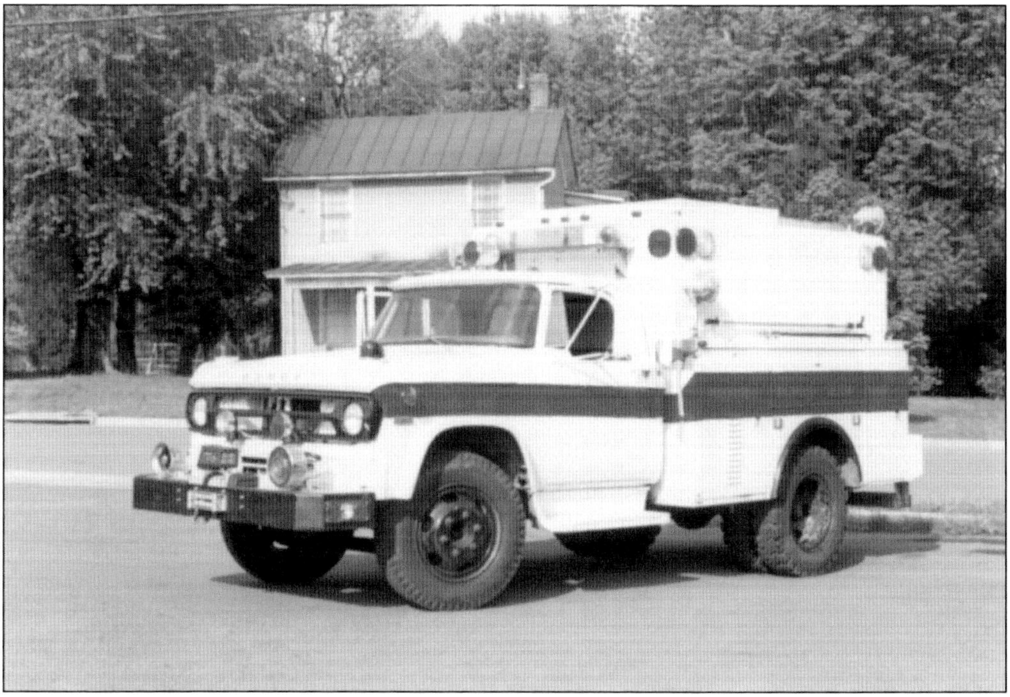

In the mid-1980s, the VEPCO power company donated this 1969 Dodge truck with a Reading utility body to the Purcellville Volunteer Rescue Squad. Members from the rescue squad equipped it to run as a squad truck, adding extrication equipment and a generator. When the truck was no longer used, the extrication equipment was given to the Purcellville Volunteer Fire Department and was placed on a fire engine.

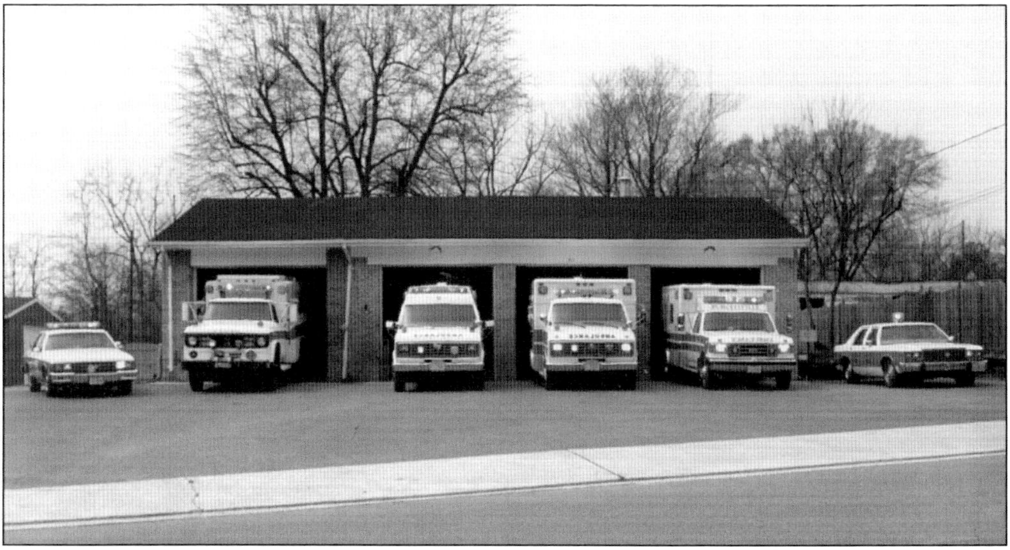

This late-1980s or early-1990s photograph shows the diversity of the apparatus roster of the Purcellville Volunteer Rescue Squad. On each side is an EMS chase vehicle, each received from the sheriff's department. The 1969 Dodge squad truck occupies the far left bay. Three different types of Ford ambulances are also represented: from left to right, a Ford van, an Econoline box ambulance, and a Ford F-350 box ambulance. (Courtesy of Purcellville VRS.)

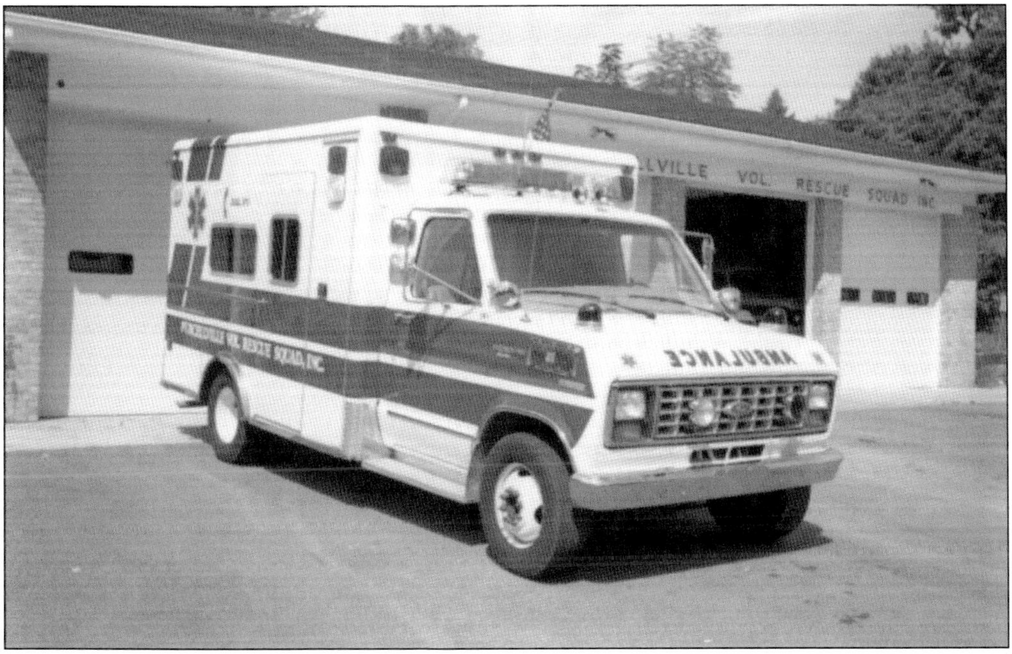

The Purcellville Volunteer Rescue Squad liked their Ford van built by Horton, so they returned to Horton for their next unit. Ambulance No. 141 is an early-1990s Ford Econoline painted white with several red stripes. This was the first box-style ambulance purchased by Purcellville's rescue squad. (Courtesy of Purcellville VRS.)

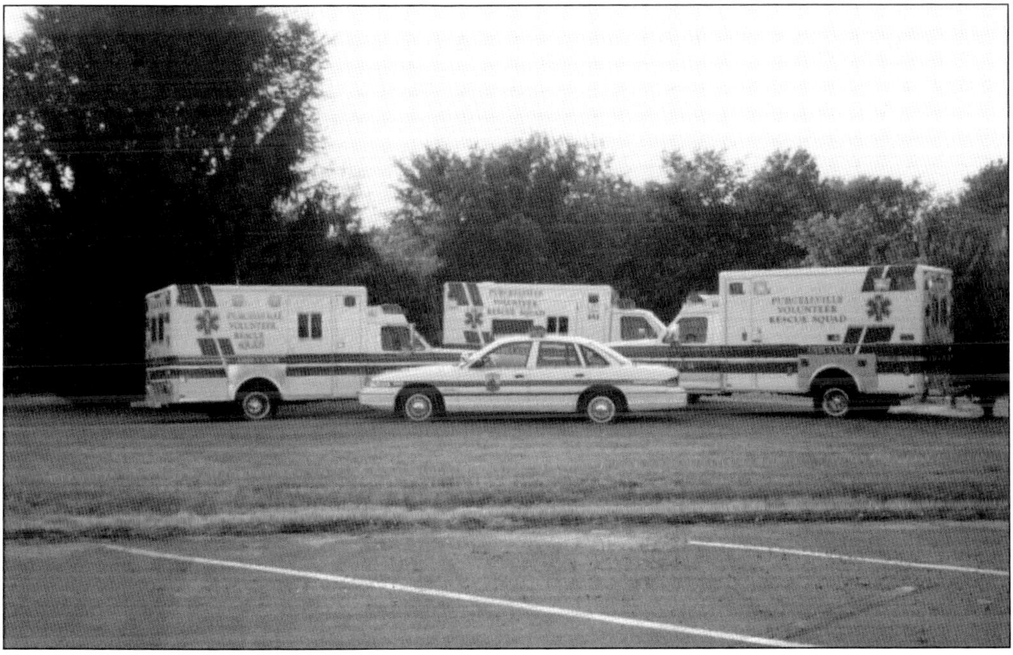

During the 1990s and after 2000, the Purcellville Volunteer Rescue Squad's fleet of ambulances has consisted of Ford F-350s built by Horton. An EMS chase vehicle (from the old sheriff's department) was photographed with the department's three Ford F-350/Hortons. (Courtesy of Purcellville VRS.)

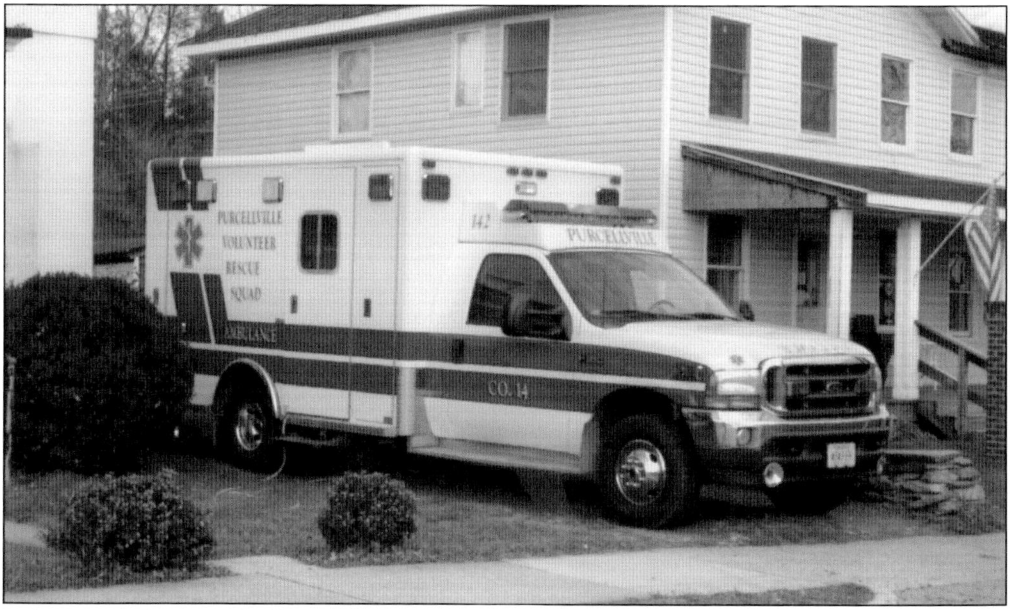

In late 2006, the Purcellville Volunteer Rescue Squad moved its operations to temporary quarters. The original station on Twentieth Street was old and cramped. Until a decision was made on a new facility to house both the fire department and rescue squad, the old station was renovated, with work being completed in early 2007. Ambulance No. 14-2 is a 2001 Ford F-350/Horton in service at the temporary facility on East Main Street in late 2006.

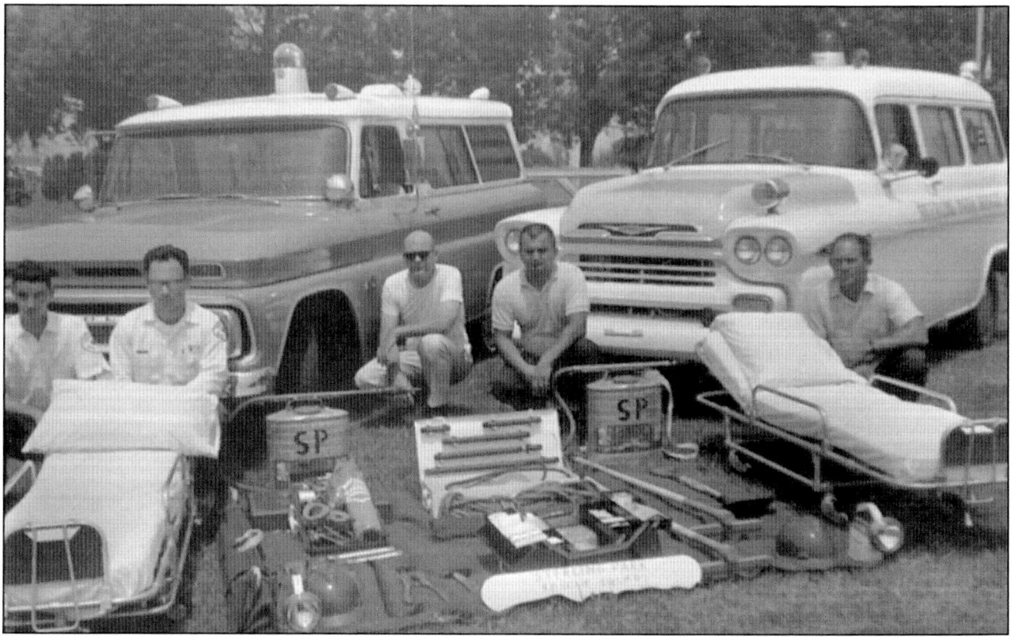

The Sterling Park Rescue Squad was organized in 1964 as the Sterling Park Jaycees' "Grand Project." The department's first vehicle was this 1959 Chevrolet Carryall (right), which was purchased from Manchester, Virginia, for $500. It was destroyed on May 7, 1968, when it caught fire. A 1966 Chevrolet Carryall (left) was the first new ambulance purchased by the department. (Courtesy of Sterling VRS.)

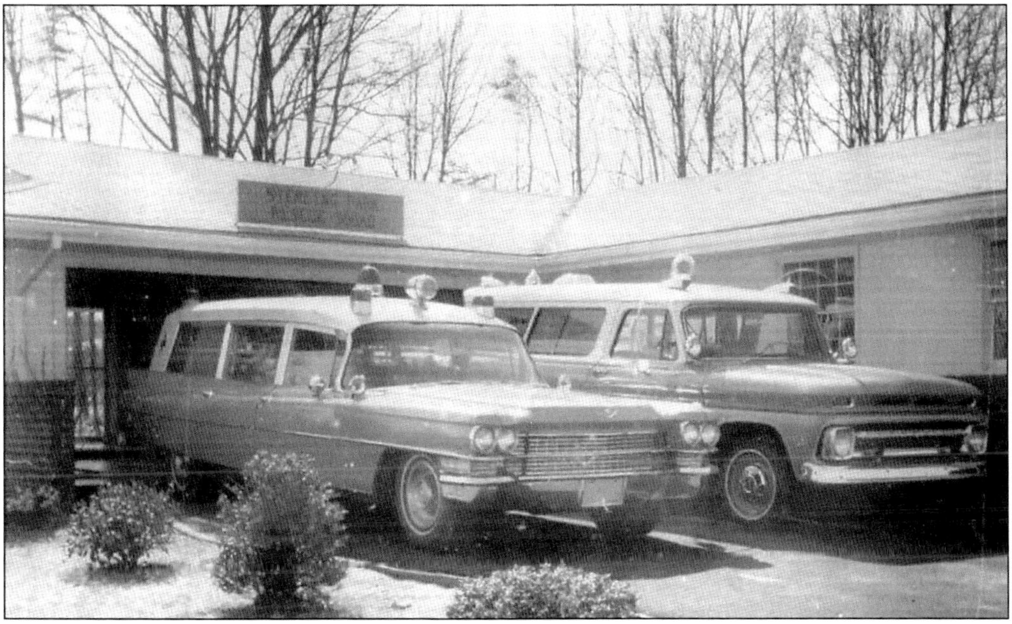

The Sterling Park Rescue Squad's first station was the library located at 304 North Sterling Boulevard. When the department received its heavy squad, it was kept at a local Shell gas station. Rescue apparatus were painted green and white. Pictured at the library is a 1963 Cadillac ambulance that was purchased from McLean, Virginia, and the 1966 Chevrolet Carryall. (Courtesy of Sterling VRS.)

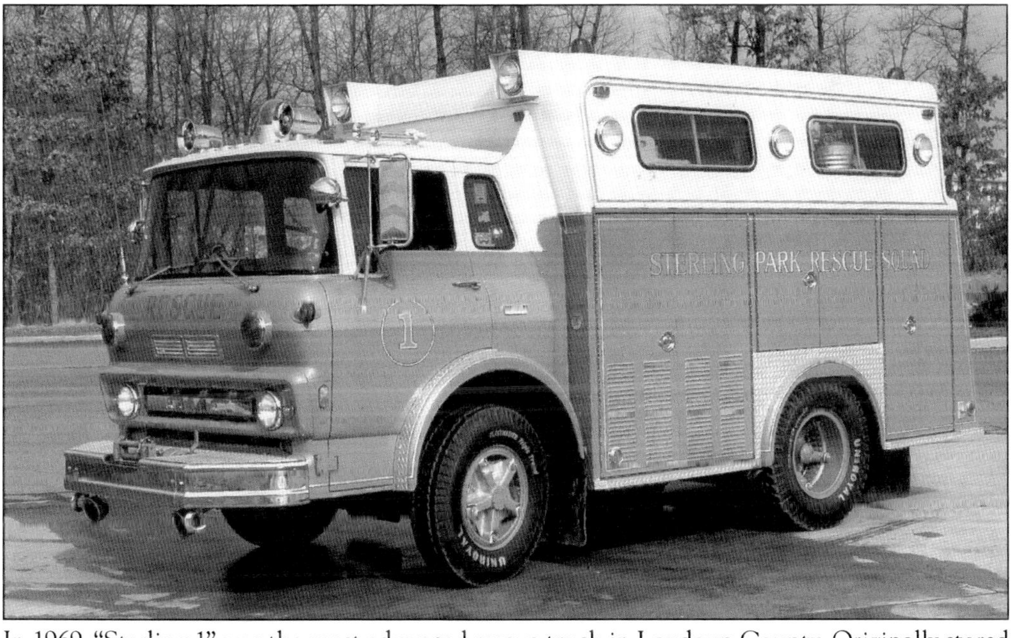

In 1969, "Sterling 1" was the most advanced rescue truck in Loudoun County. Originally stored at the Shell gas station, it was photographed on the front ramp of Fire Station No. 11 and Rescue Squad No. 15, located at 104 East Commerce Street, where they are located today. After service in Sterling Park, the 1969 GMC/Swab was obtained by the Stafford County, Virginia, sheriff's department, painted brown, and used as a SWAT vehicle. (Courtesy of Robert Kimball.)

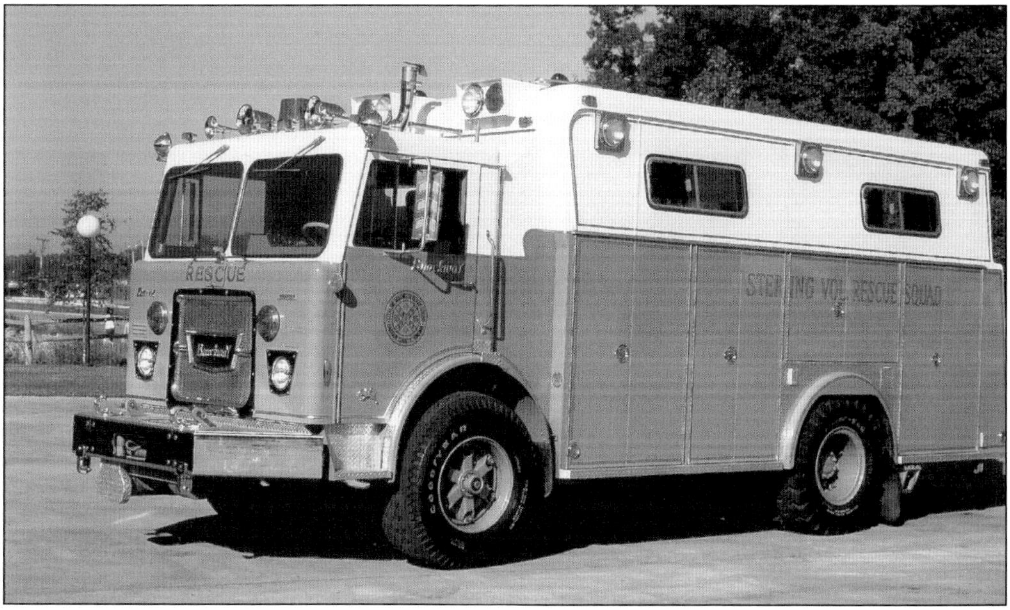

The Sterling Rescue Squad purchased a new heavy squad in 1973, when they placed into service this unusual Brockway with a rescue body built by Swab. Originally painted green and white (as shown above), it was later repainted white with wide green stripes. The Brockway cab would later be replaced. By this time, apparatus were labeled "Sterling" instead of "Sterling Park." (Courtesy of Chuck Madderom.)

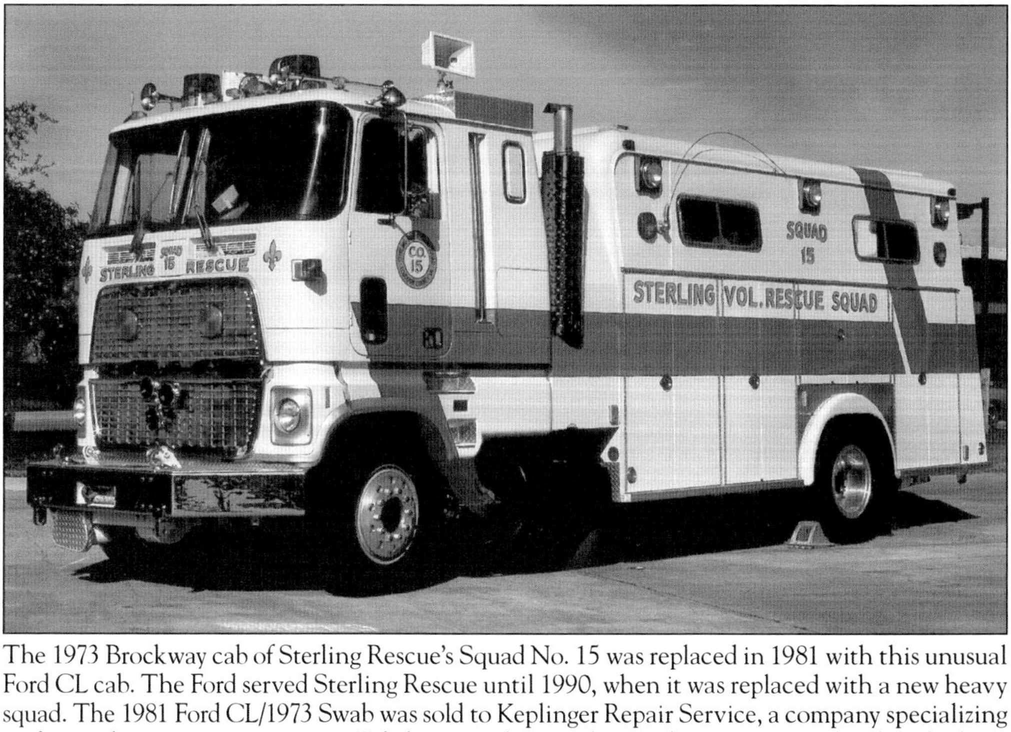

The 1973 Brockway cab of Sterling Rescue's Squad No. 15 was replaced in 1981 with this unusual Ford CL cab. The Ford served Sterling Rescue until 1990, when it was replaced with a new heavy squad. The 1981 Ford CL/1973 Swab was sold to Keplinger Repair Service, a company specializing in fire and rescue maintenance. While at Keplinger, the truck was repainted red and white. (Courtesy of Robert Kimball.)

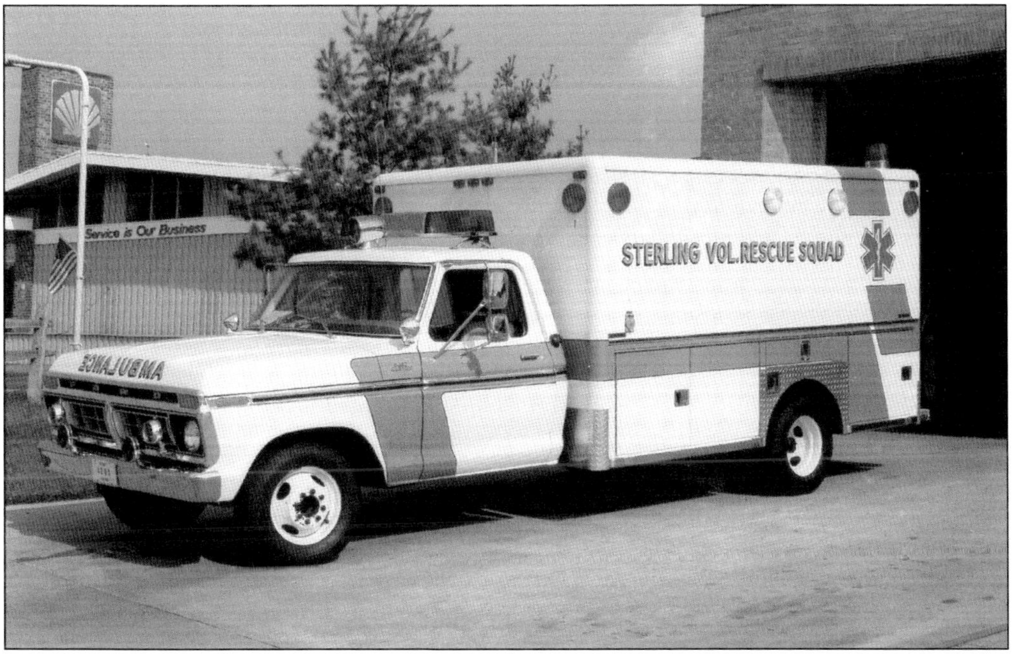

Swab ambulances were a popular choice with many fire departments and rescue squads. In 1978, Sterling Rescue purchased a Swab ambulance mounted on a Ford F-350. Photographed on the ramp of the rescue station on East Commerce Street, the ambulance was painted white with green stripes. (Courtesy of Chuck Madderom.)

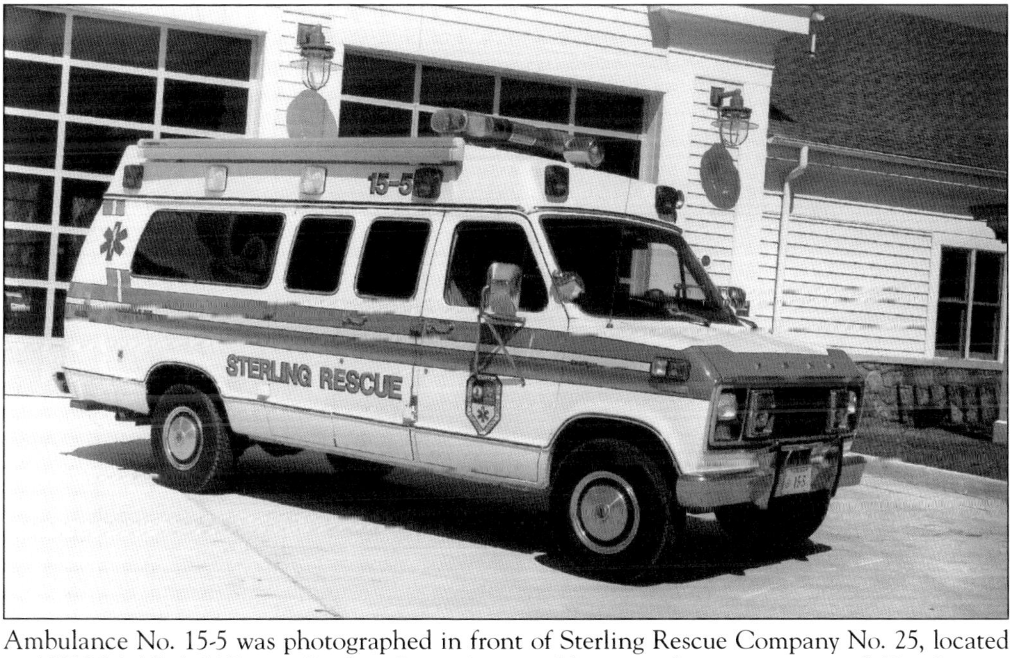

Ambulance No. 15-5 was photographed in front of Sterling Rescue Company No. 25, located at 46700 Middlefield Drive. Sterling Rescue's second location shares quarters with Fire Station No. 18. Wheeled Coach built the ambulance on a 1979 Ford Econoline 350. The ambulance was later converted to a rehab unit, which included the addition of an awning. The awning was installed along the ambulance's roofline, as shown above.

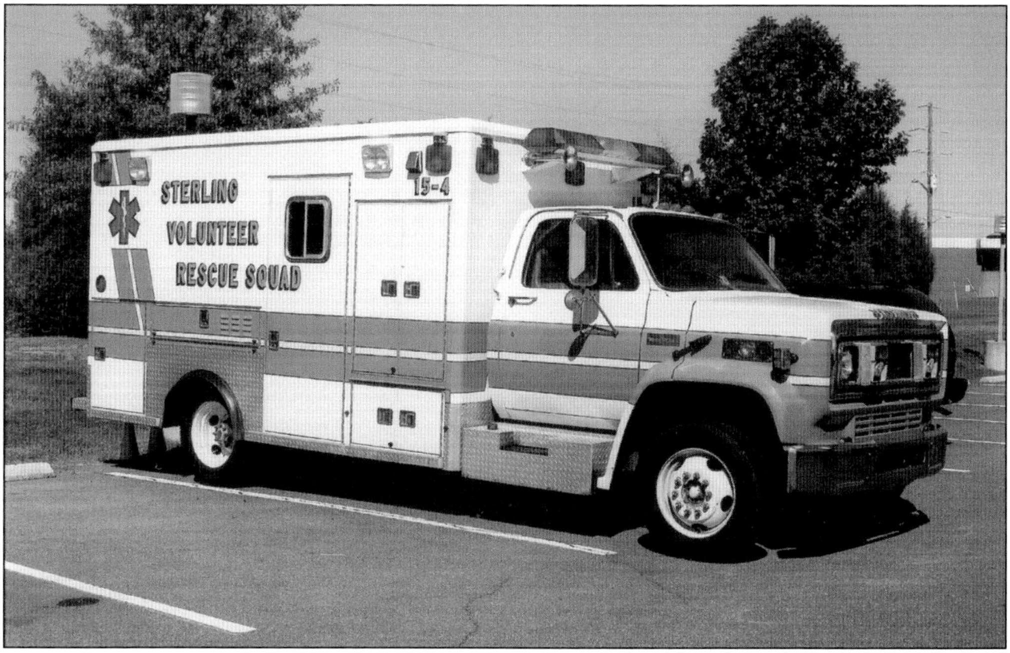

Sterling Rescue purchased two similar monster medics, the first such ambulances in Loudoun County. Monster medics are ambulances with a larger box and truck chassis. Medic No. 15-4 was a 1989 Chevrolet 70 built by Wheeled Coach. This ambulance would later be used by the fire marshal's office as a bomb squad. The other ambulance was sold to the Lucketts VFD.

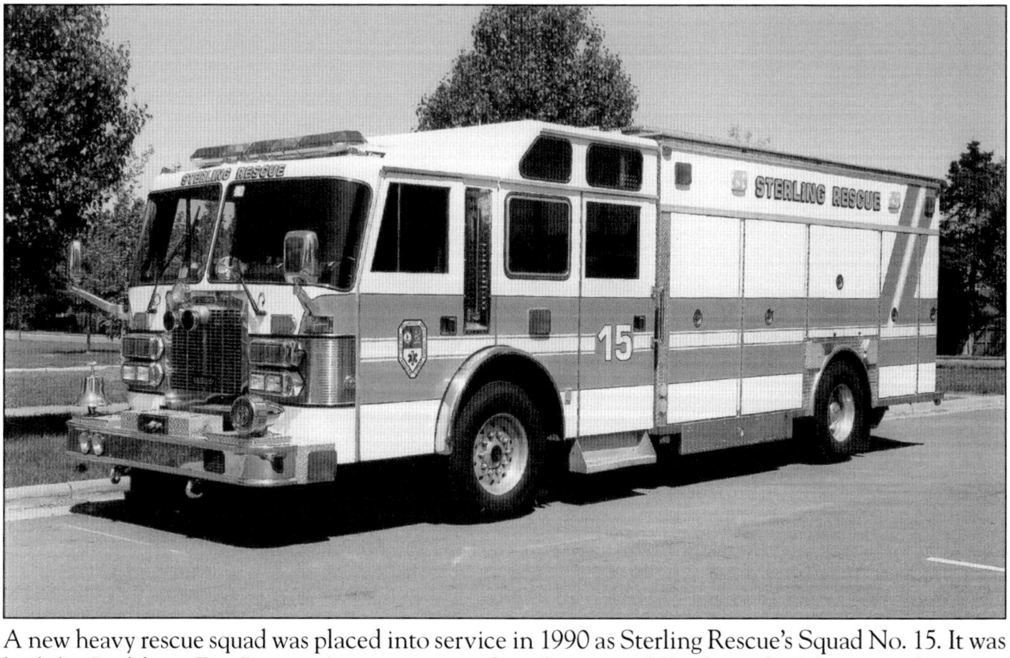

A new heavy rescue squad was placed into service in 1990 as Sterling Rescue's Squad No. 15. It was built by Saulsbury Fire Rescue Apparatus on a four-door raised Duplex cab. After providing over 10 years of service, the truck was returned to Saulsbury in Tully, New York, to be reconditioned for extended service to Sterling. Squad No. 15 was photographed just after returning from Saulsbury in 2001.

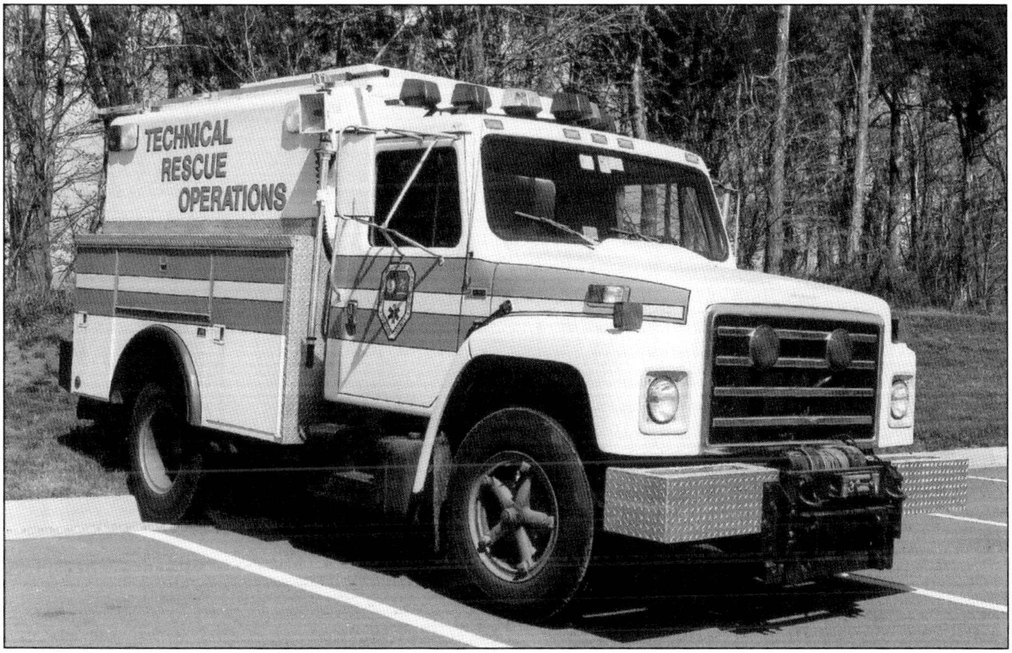

The Sterling Rescue Squad is part of a state disaster-response unit called Task Force No. 8. It is deployed at the request of the governor or the state to provide assistance to parts of Virginia affected by disasters. Technical Rescue No. 25 (Tech No. 25) is a 1983 International that was once a power-company truck. It carries a variety of specialized equipment for technical-rescue emergencies.

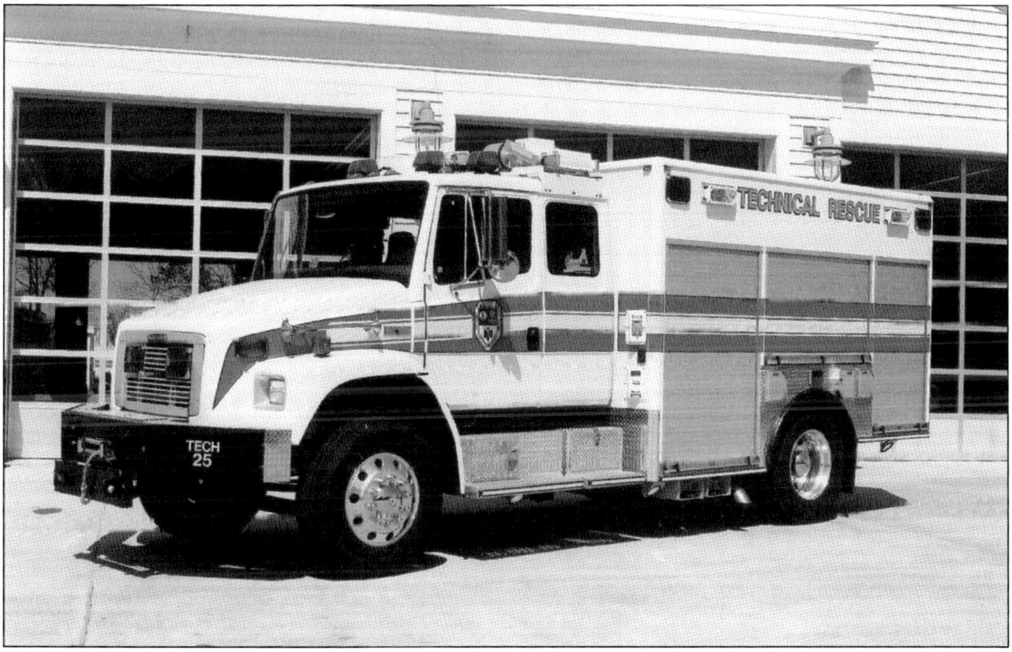

Sterling Rescue's 1983 technical-rescue truck was replaced in 2005 by this American LaFrance built on a Freightliner. Tech No. 25 carries a variety of specialized equipment and has roll-up doors, a winch, and a light tower. A unique feature is that the cab has three bucket seats—two in front and one in the rear. Tech No. 25 also operates as a backup to Squad No. 15.

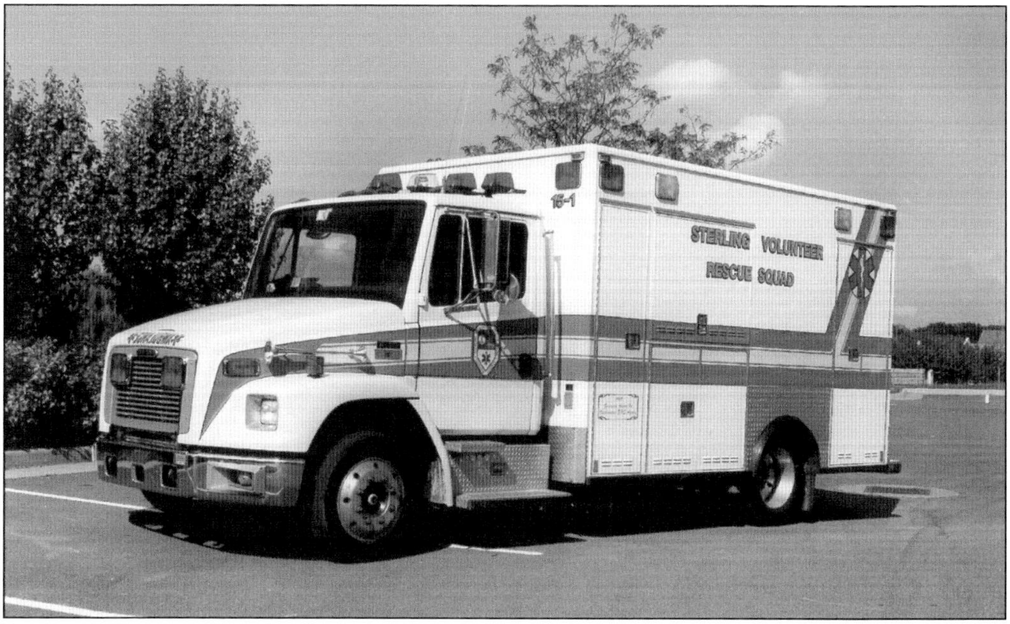

Sterling Rescue prefers the larger monster medic built on a Freightliner cab. The first units were purchased in 1995 and were built by Medic Master. The squad continues to purchase them today. Ambulance No. 15-1, a 1995 ambulance, served Sterling until 2005. It was donated to Hancock County, Mississippi, after Hurricane Katrina, and it is used as a command and tactical operations unit.

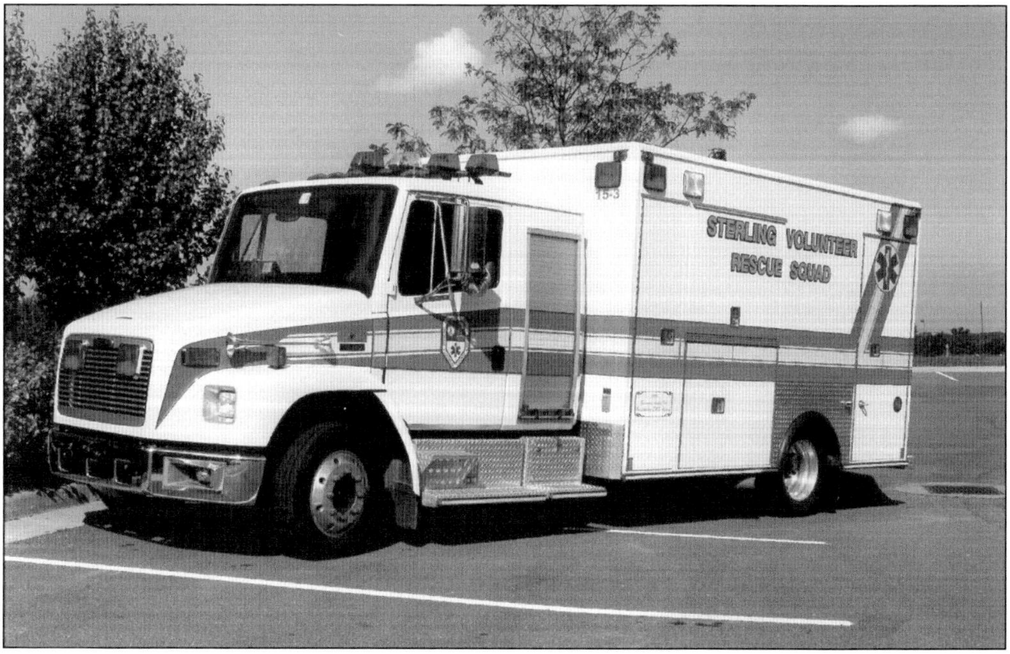

This Sterling Rescue 1999 Freightliner/Medic Master ambulance has one added feature. It was also designed as a rehabilitation unit. The larger compartment with a roll-up door located behind the driver carries additional supplies for rehab work required at larger emergency incidents. Ambulance No. 15-3 also has an awning built on to the ambulance body.

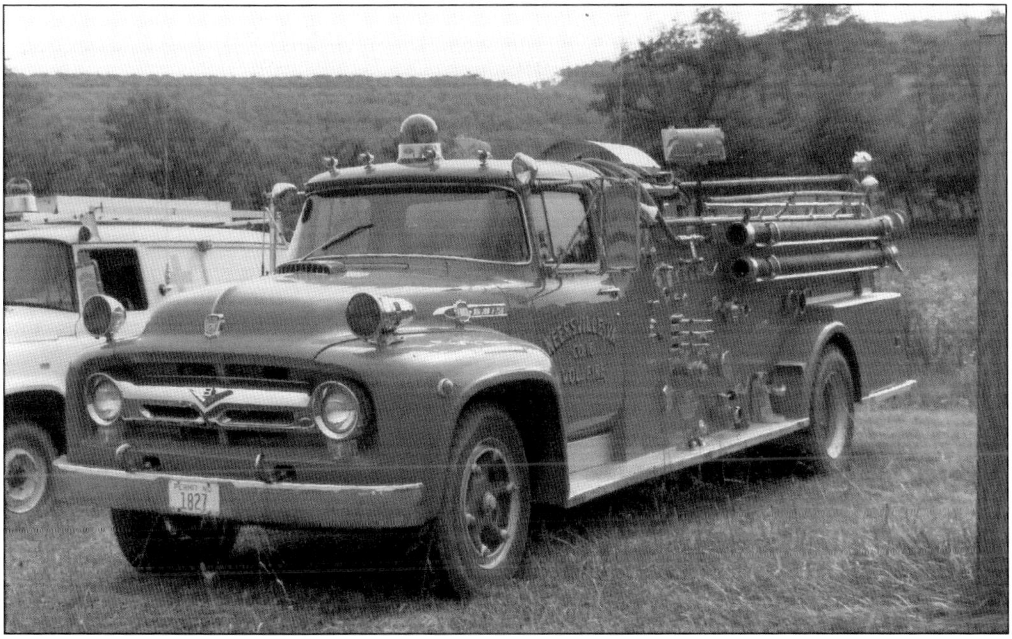

The Neersville Volunteer Fire Department was established as a substation to the Round Hill VFD Station No. 4 in 1976. An ambulance was the first unit placed in service and was housed in a garage in the Loudoun Heights area. The first pumper was this 1956 Ford Big Job built by American. It was originally ordered by Hamilton, sold to Round Hill, and then donated to Neersville. (Courtesy of Dale Fletcher.)

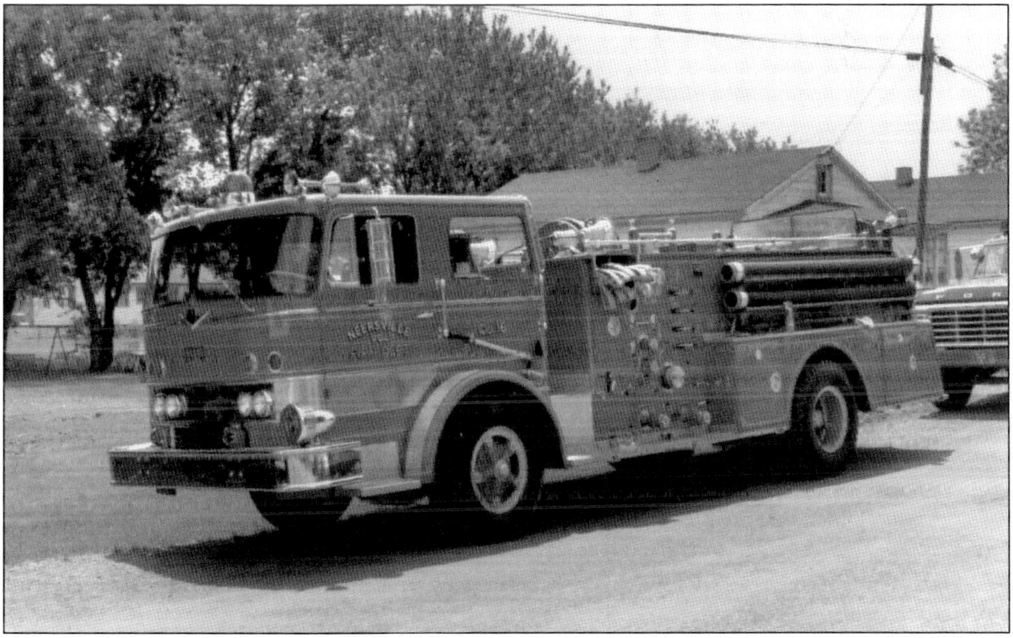

Neersville has purchased a variety of used fire and rescue apparatus from other Loudoun County stations. In 1984, the volunteers purchased this 1967 International CO190/Oren fire truck from the Ashburn Volunteer Fire Department. The International was equipped with a 1,000-gallon-per-minute pump and carried 600 gallons of water. It was replaced in 1989.

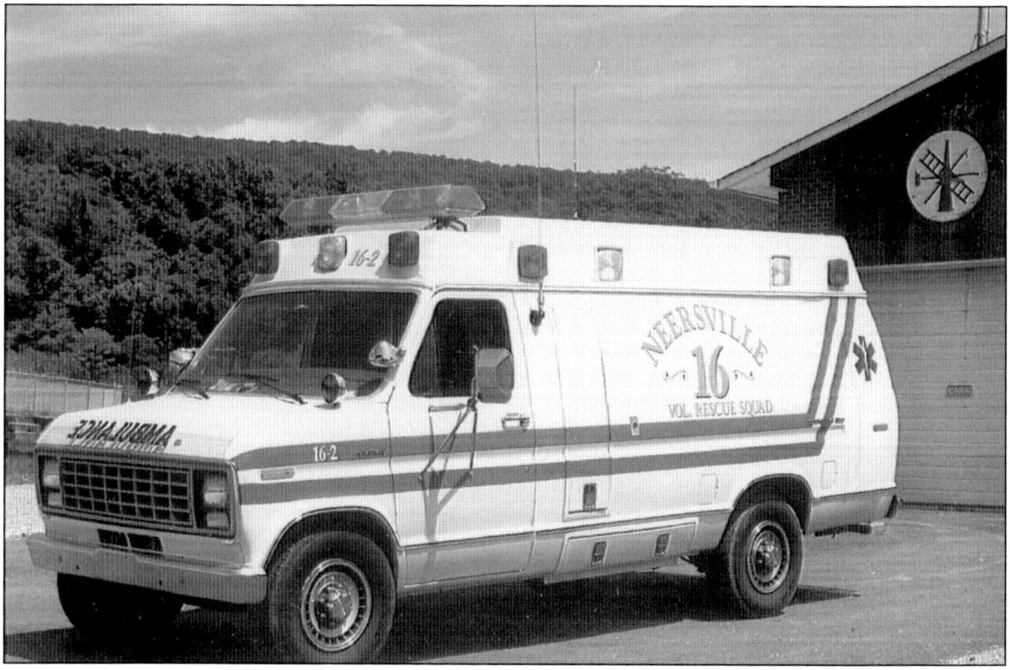

Designated as Ambulance No. 16-2, this 1980s Ford Econoline ambulance was painted white with a red stripe and provided basic life support response to the Neersville community. The Neersville VFD made the difficult decision in 2003 to discontinue ambulance service and provide first-responder service through the fire department. Neersville uses the help and support of other communities to provide ambulance transportation to the hospital.

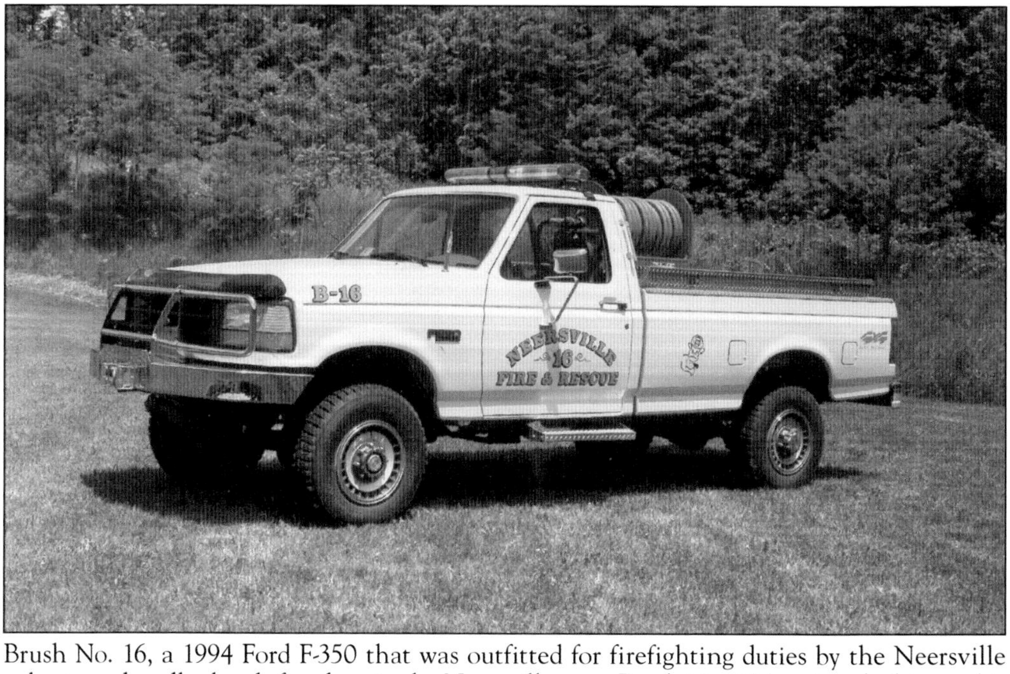

Brush No. 16, a 1994 Ford F-350 that was outfitted for firefighting duties by the Neersville volunteers, handles brush-fire duty in the Neersville area. Brush No. 16 is painted white and is equipped with a 150-gallon water tank.

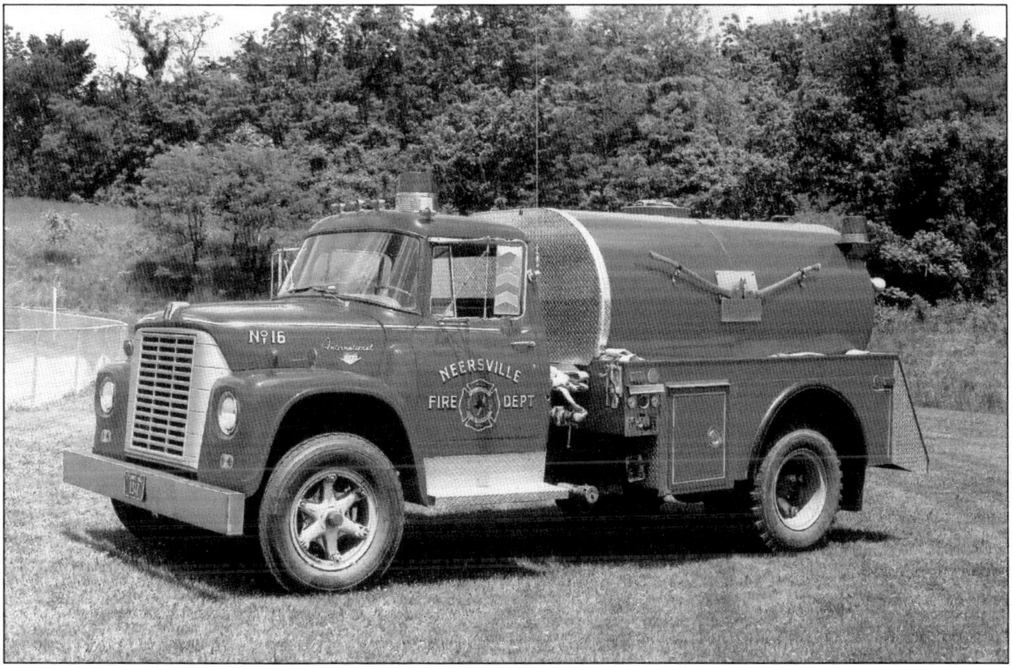

In 1997, Neersville Station No. 16 added a much-needed fire truck to its apparatus roster. Neersville's first tanker truck was this 1967 International, which has a 1980 4-Guys 1,500-gallon tanker body. Neersville purchased this tanker from the Philomont VFD, Station No. 8 in Loudoun County. It served until 2001, when Neersville took delivery of a new tanker truck.

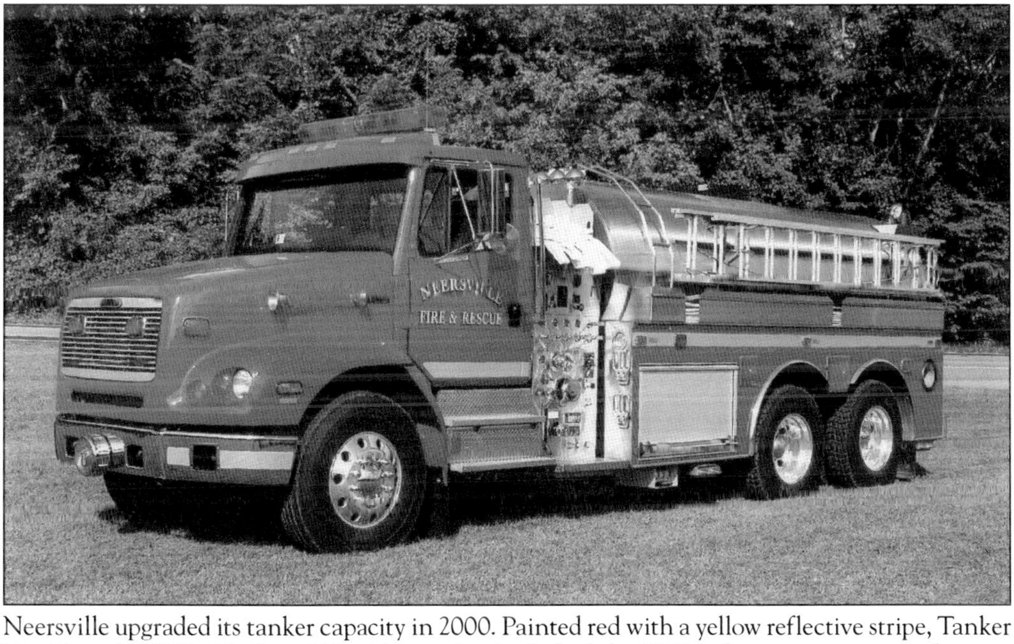

Neersville upgraded its tanker capacity in 2000. Painted red with a yellow reflective stripe, Tanker No. 16 is a 2000 Freightliner 112 with a tanker body built by S&S, a fire-apparatus manufacturer in Fairmount, Indiana, that specializes in building tanker trucks. Equipped with a 750-gallon-per-minute pump and carrying 3,000 gallons of water, Tanker No. 16 also carries additional hose and attack lines.

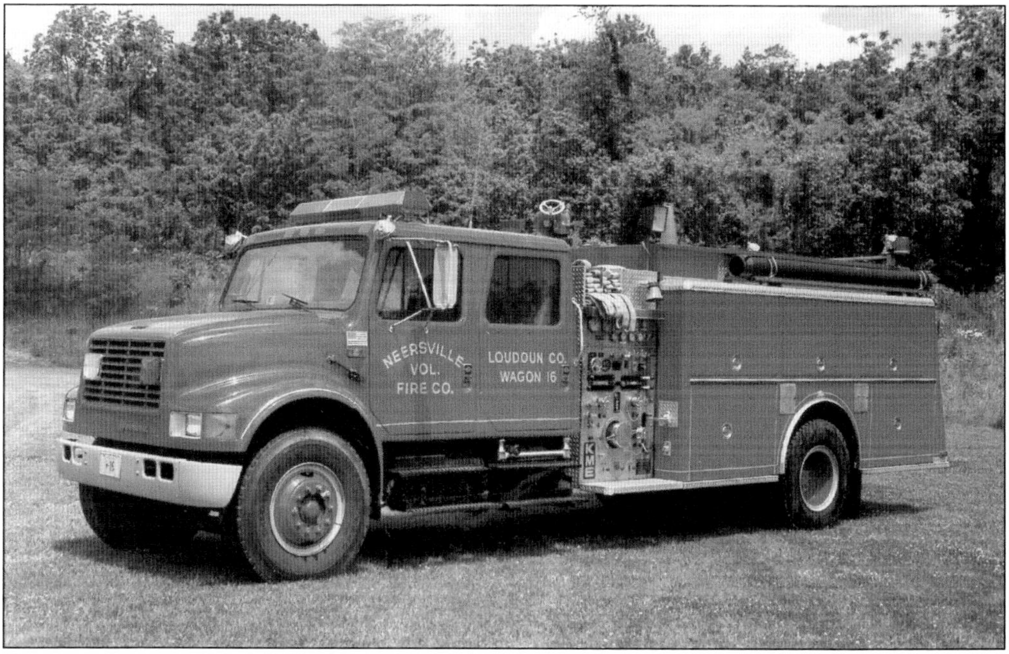

The first new fire truck purchased by Neersville was Wagon No. 16, a red 1989 four-door International built by KME of Pennsylvania. The new Wagon No. 16 came equipped with a larger 1,000-gallon water tank because of the lack of hydrants in the Neersville area. In 2006, Wagon No. 16 was given to the Loudoun County Fire Rescue Training Center.

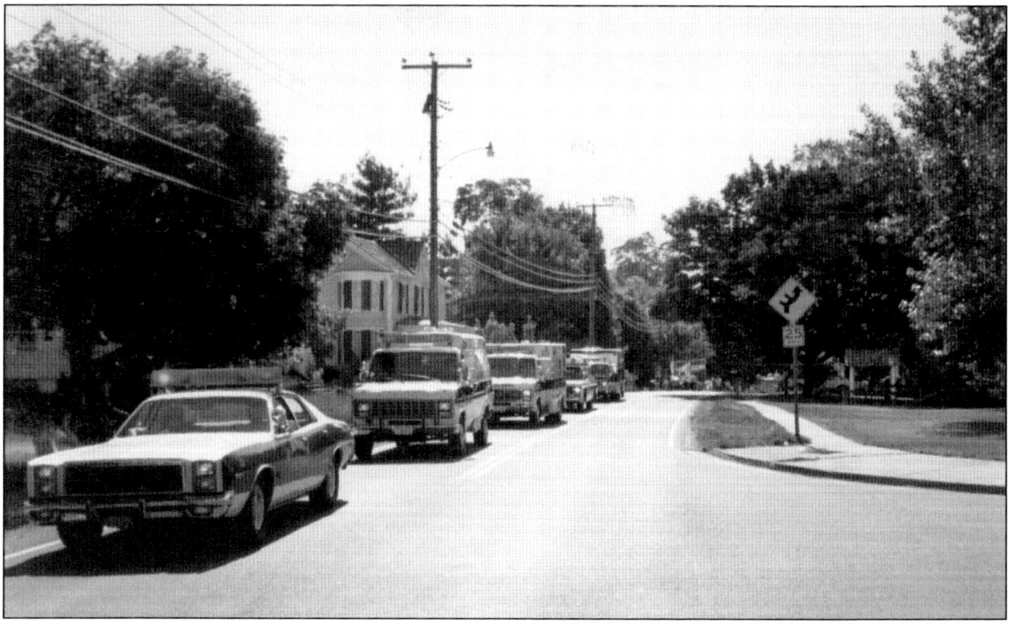

The Loudoun County Rescue Squad was formed in 1952 and was first established in Hamilton. When a second squad truck was placed in Leesburg, several members decided to establish their own independent rescue squad in Hamilton. In 1979, the Hamilton Volunteer Rescue Squad was created with two ambulances and a boat. This parade photograph shows some of the early Hamilton rescue apparatus. (Courtesy of Hamilton VRS.)

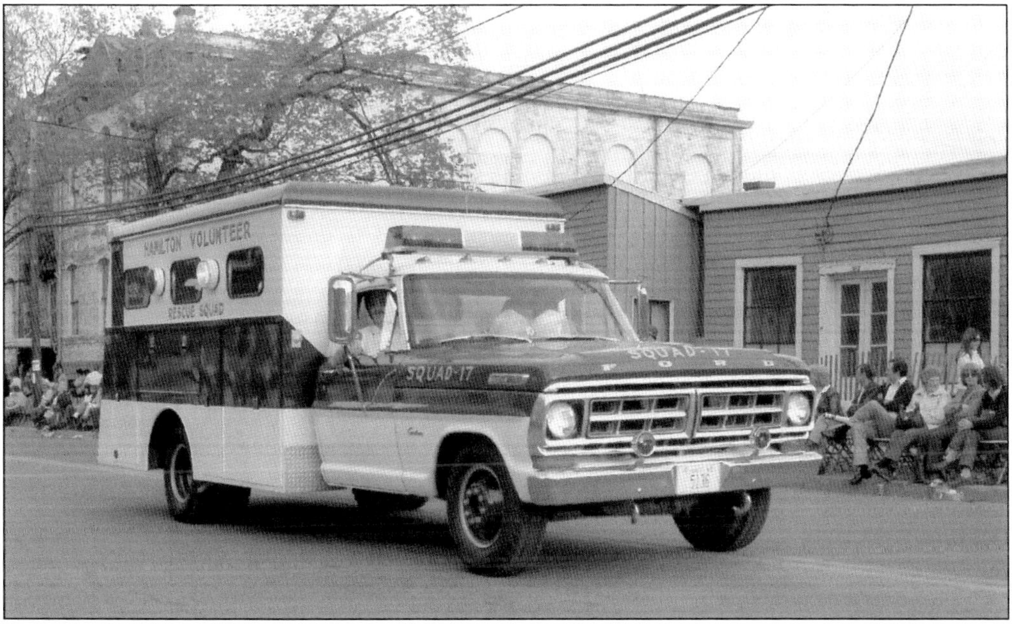

In 1982, the Hamilton Rescue Squad purchased this 1973 Ford F-350 with a squad box built by Reading. The Rouss Fire Company of Winchester, Virginia, originally owned the truck, and it was painted a dark green and white. Hamilton repainted the squad truck its colors, white and green. This would be the first of two squad trucks purchased from Winchester departments and would later be rebuilt. (Courtesy of Dave Bowen.)

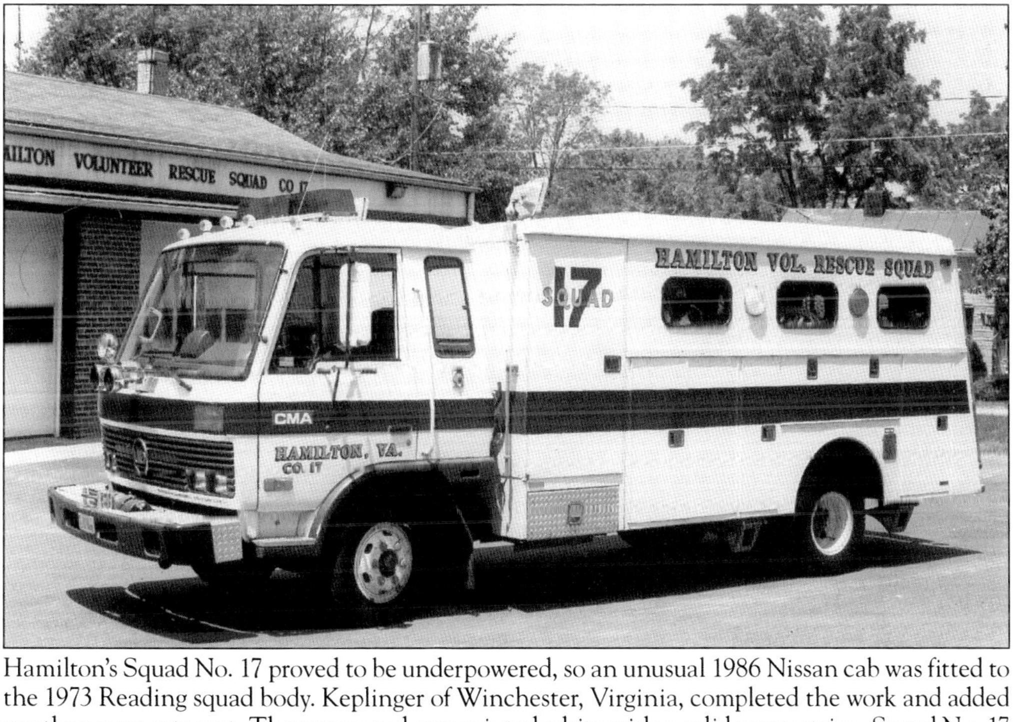

Hamilton's Squad No. 17 proved to be underpowered, so an unusual 1986 Nissan cab was fitted to the 1973 Reading squad body. Keplinger of Winchester, Virginia, completed the work and added another compartment. The new squad was painted white with a solid green stripe. Squad No. 17 was photographed in front of the old rescue squad station on North Laycock Street in 1998.

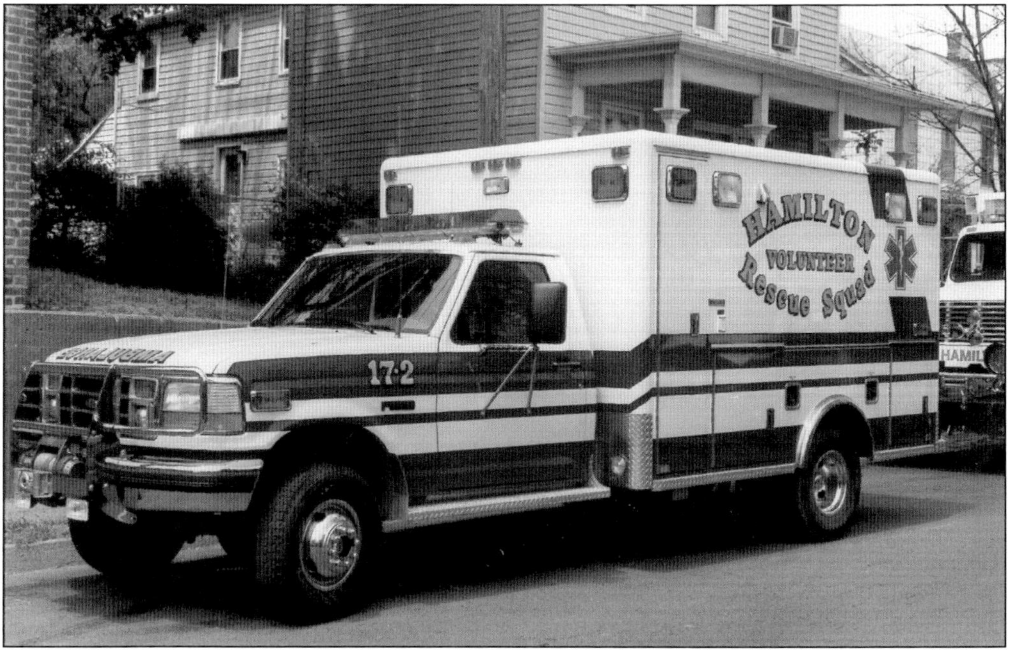

Since separating from the Loudoun County Rescue Squad, the Hamilton Volunteer Rescue Squad has used a variety of ambulances built by Horton on a Ford cab and chassis. Ambulance No. 17-2 is a 1994 Ford/Horton painted white with green stripes. It was photographed at the 2000 Apple Blossom Festival parade in Winchester.

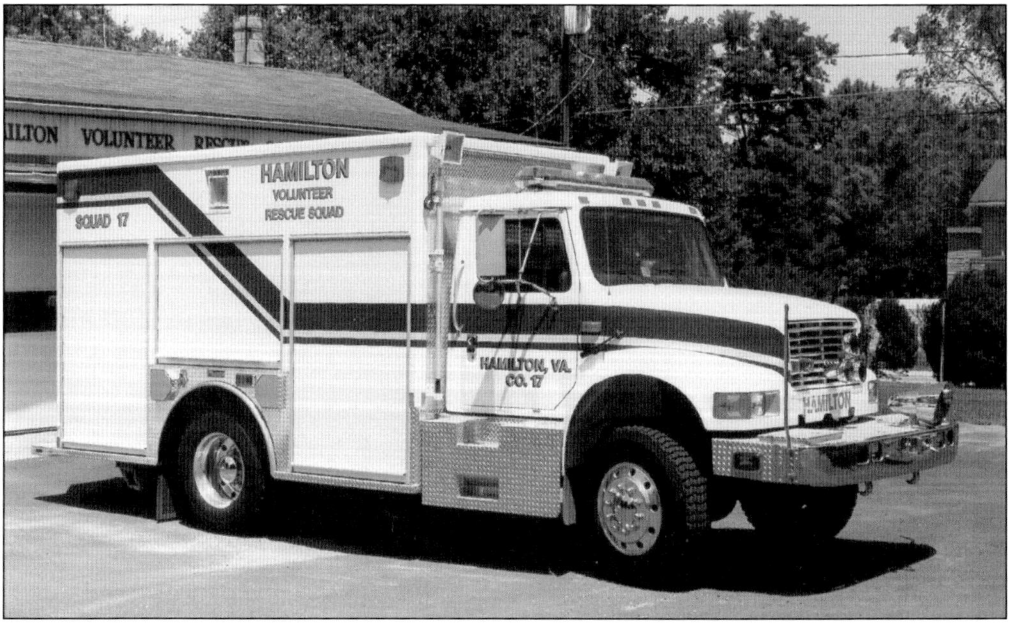

When it came time to replace the Nissan squad in the late 1990s, Hamilton Rescue went back to a Winchester department to purchase a 1994 International with a squad body built by Marion. The International previously ran with the Winchester Volunteer Rescue Squad (now the Millwood Station Fire-Rescue). The squad box has compartments only, so additional manpower must respond in Hamilton's Special Emergency Response Vehicles (SERV) or utility vehicles.

Three

LOUDOUN COUNTY DEPARTMENT OF FIRE, RESCUE, AND EMERGENCY MANAGEMENT

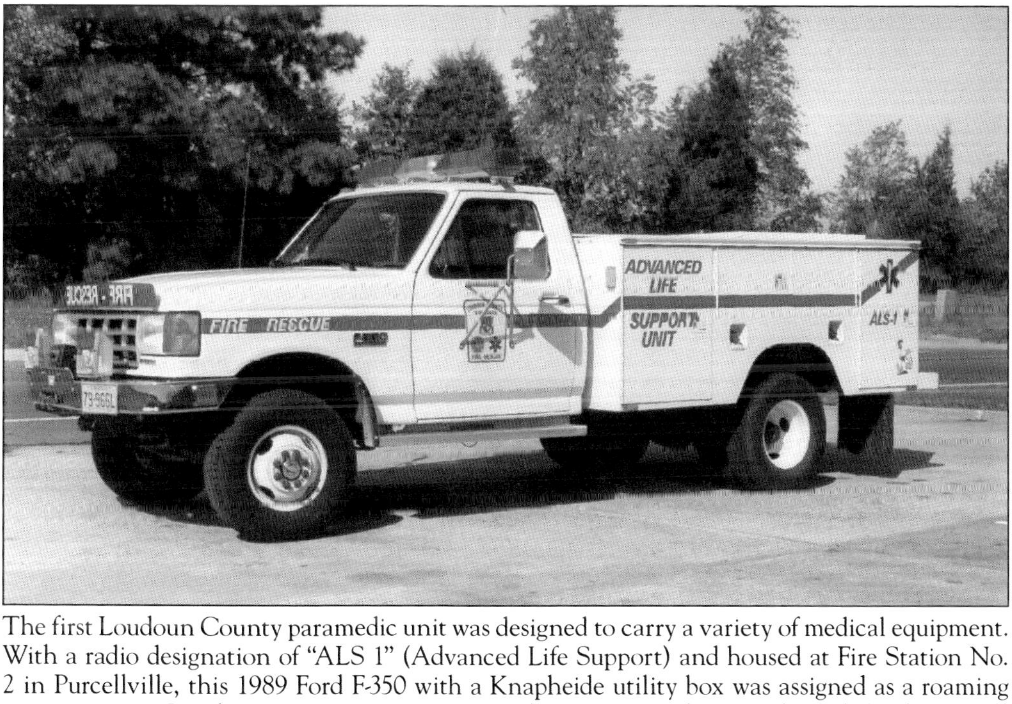

The first Loudoun County paramedic unit was designed to carry a variety of medical equipment. With a radio designation of "ALS 1" (Advanced Life Support) and housed at Fire Station No. 2 in Purcellville, this 1989 Ford F-350 with a Knapheide utility box was assigned as a roaming unit in western Loudoun County. ALS No. 1 was later reassigned as a work truck for the county sign shop.

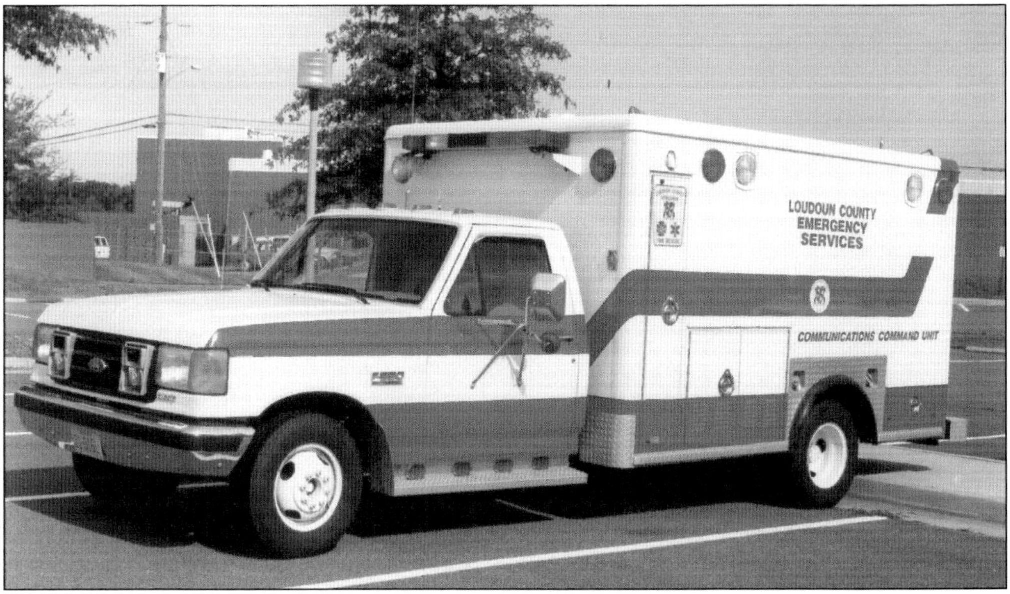

Loudoun County's first communications command unit was Command No. 50. It was an older ambulance converted into a usable field unit by Loudoun County fire personnel. The ambulance box is from an old Fairfax County, Virginia, ambulance that was obtained by Station No. 13, the Loudoun County Rescue Squad. After serving at Station No. 13, where it was upgraded with a newer 1987 Ford F chassis, the ambulance was donated to the county.

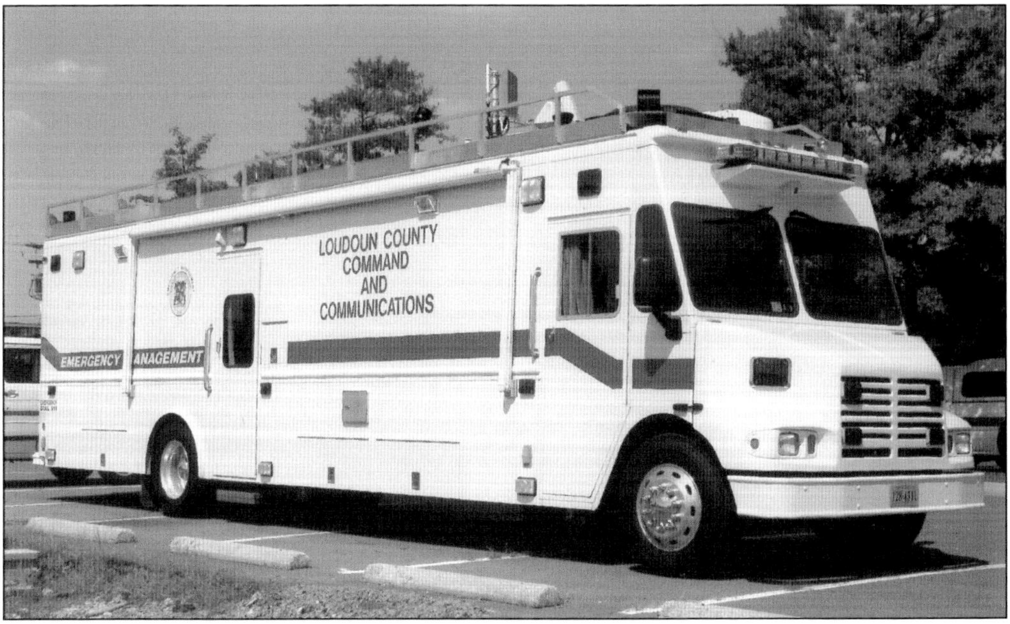

The old Loudoun County Ford ambulance communications and command unit was replaced in 2005 with a state-of-the-art vehicle. This 2005 Freightliner MT55 was built by LVI and outfitted by Bickford Vehicles of Chantilly, Virginia. The truck has four communication consoles, a media-relations console, an onboard weather system, a command/conference area, a mast-mounted camera, and an observation platform. This truck is designed to support fire and rescue personnel during major incidents.

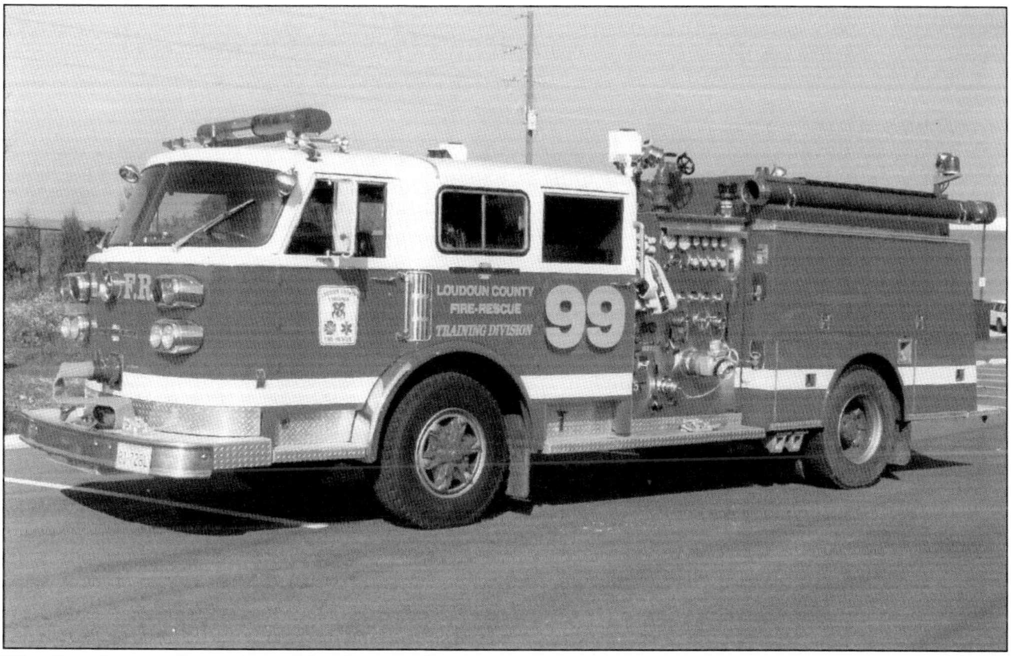

The first fire truck purchased by the Loudoun County government had an interesting background. This 1982 American LaFrance was originally ordered by the fire department of New York City (FDNY). After it served at FDNY, J. C. Moore Industries rehabbed the engine and sold it to Loudoun County, where it served as Wagon No. 99, a training engine.

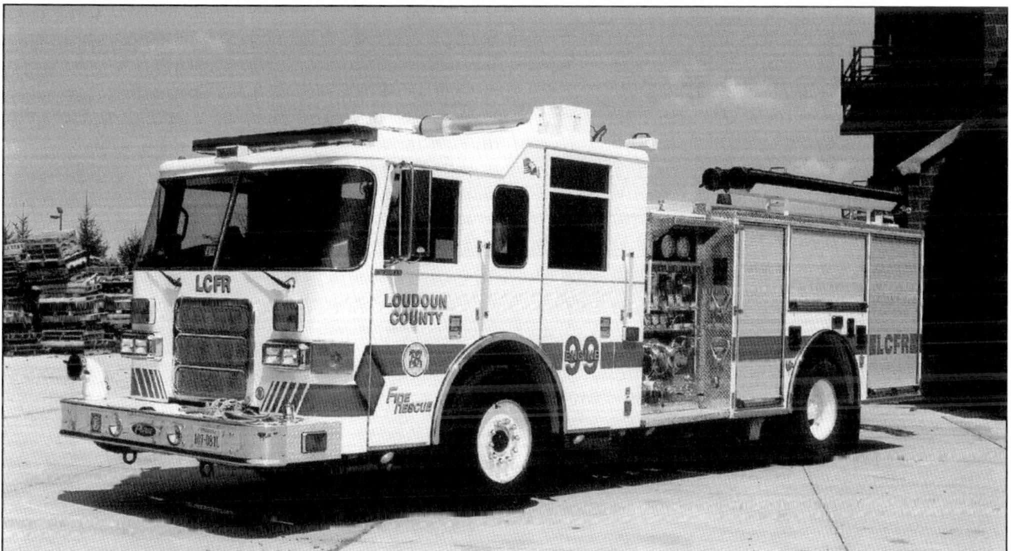

In 2001, the Loudoun County Fire Rescue Training Center received a brand-new engine for training use. This Pierce Enforcer has a 1,250-gallon-per-minute pump and carries 500 gallons of water. Interestingly the pump panel is clear so that students can have a better understanding of how the pump works by seeing the plumbing and pipes located behind the pump panel. Engine No. 99 was reassigned to Station No. 16 in Neersville.

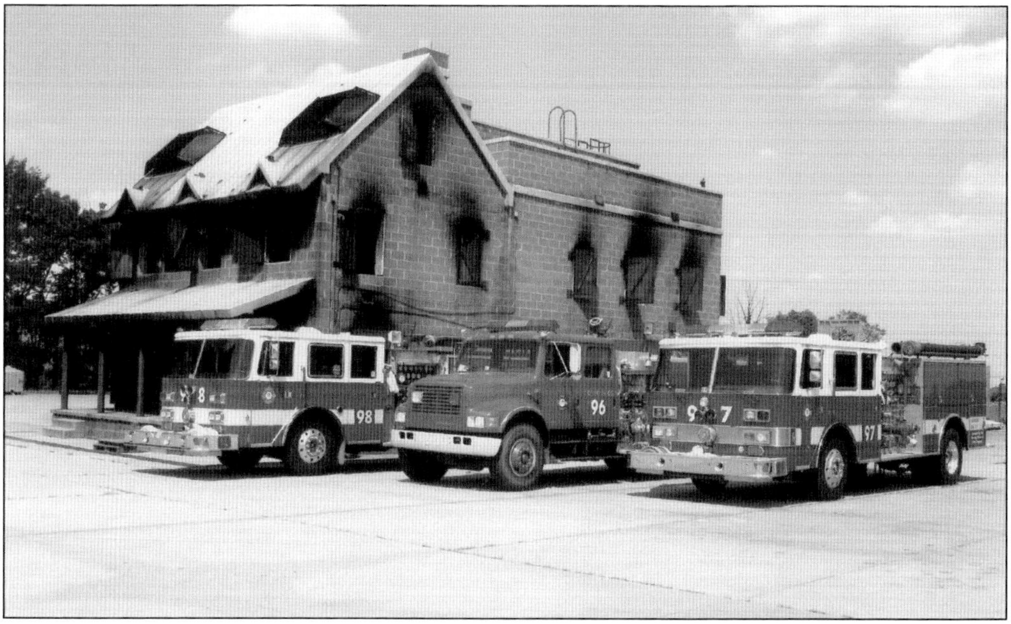

During 2005–2006, three fire trucks were assigned to the Loudoun County Fire Rescue Training Center. Engine No. 96 (center) is the 1989 International that originally saw service in Neersville. Sterling Park donated two Pierce Arrow fire trucks to the county. Engine No. 97 (right) is a 1984 model, and Engine No. 98 (left) is a 1985 model. This photograph was taken in front of the burn house at the training center.

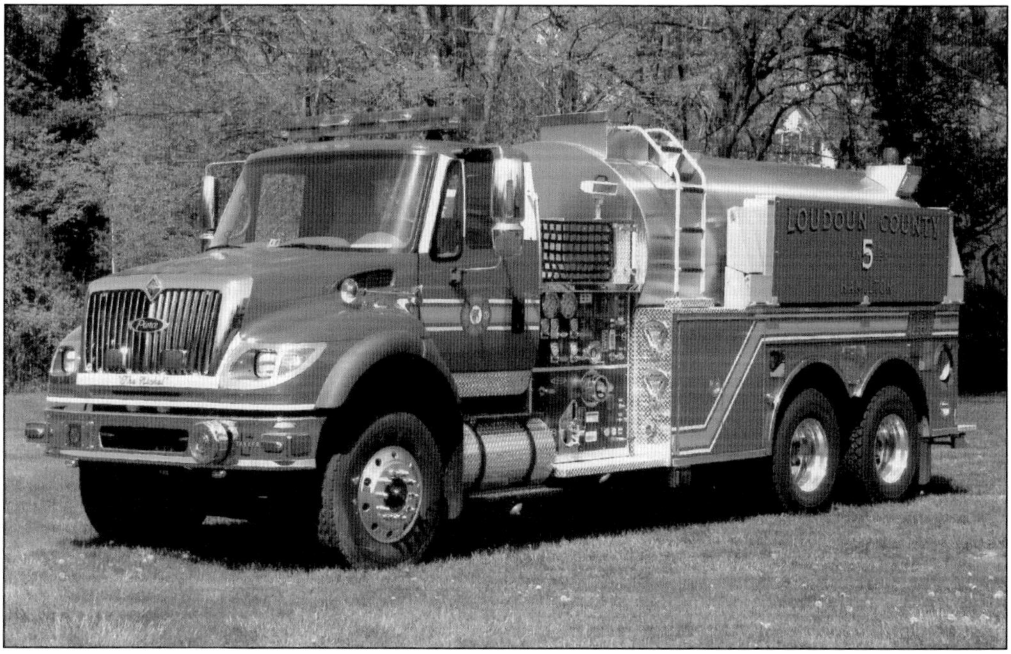

In 2005, Loudoun County took delivery of a new International tanker built by Pierce. Equipped with a 3,000-gallon water tank, it was assigned to Hamilton as Tanker No. 5. The International was one of the first county fire trucks to be painted in the new county colors of red and white with an orange stripe. Tanker No. 5 was the first of a group of new tankers purchased by the county.

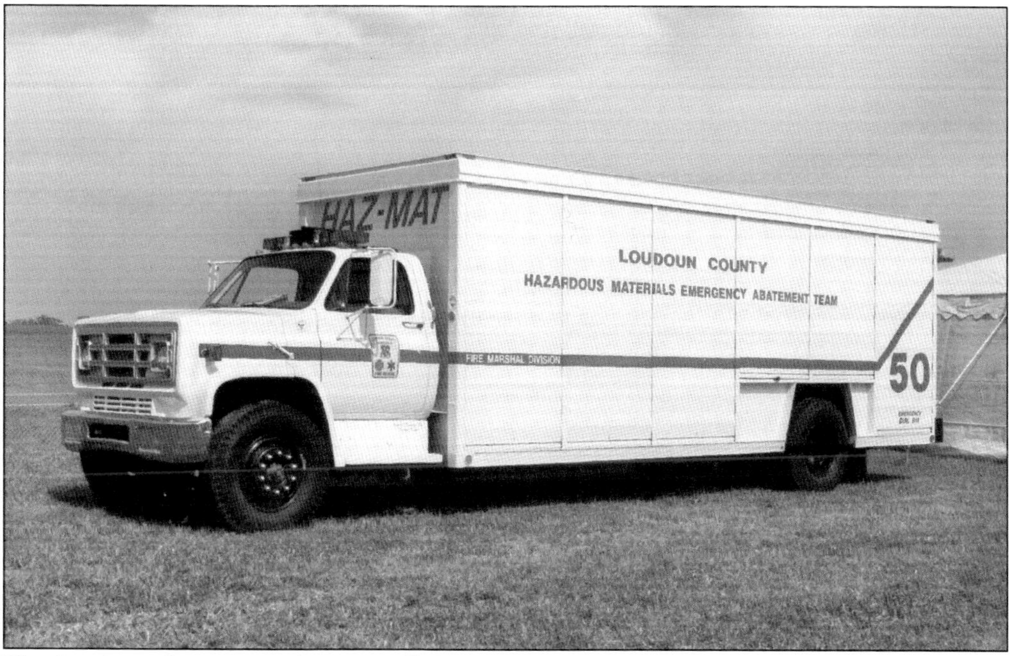

The first hazardous-materials unit staffed by Loudoun County personnel at the training center began operations around 1990. HAZMAT No. 50 was a 1978 GMC/Betten that was an old beer truck donated to the department by King Wholesale, a local beverage distributor.

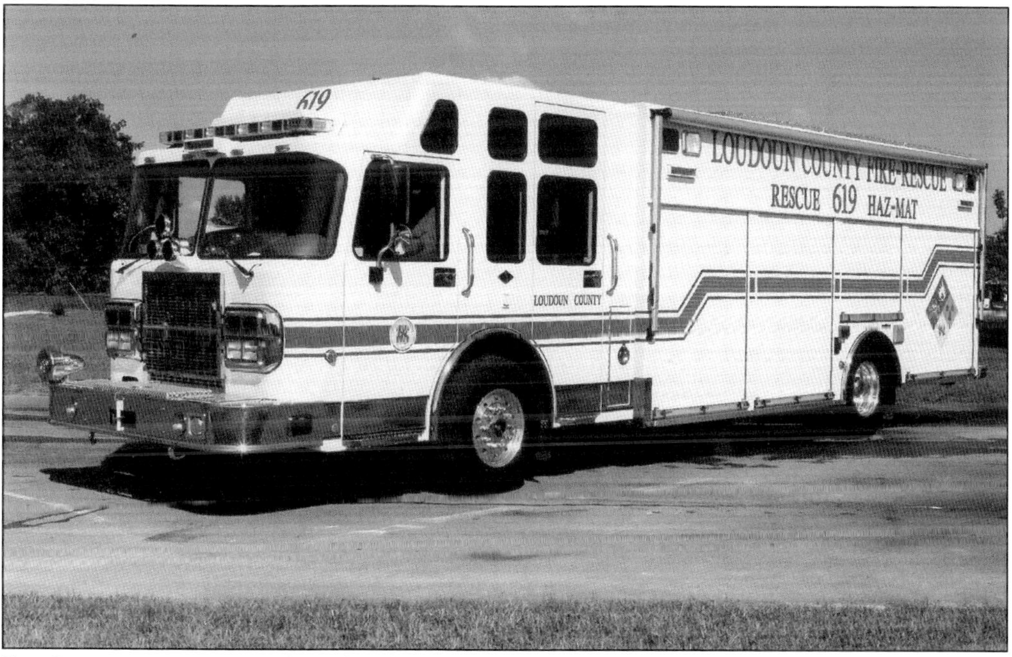

Fire Station No. 19 in South Riding houses the county HAZMAT unit. A new HAZMAT unit was delivered in 2004 and is a Spartan Gladiator built by Super Vac, Inc. HAZMAT No. 19 carries a variety of equipment and tools needed for a hazardous-materials emergency, as well as a library of reference materials. Supporting HAZMAT No. 19 are two support units assigned to the South Riding Station and Leesburg Fire Station No. 20.

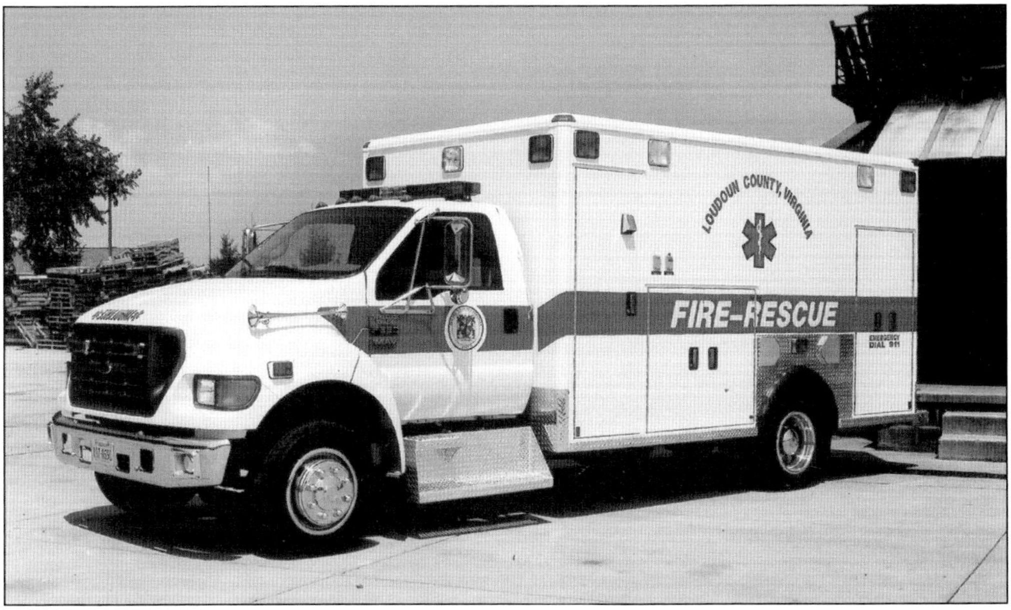

The first ambulance purchased by the Loudoun County government was this 2002 Ford F-650 built by Wheeled Coach. Assigned to the Loudoun County Fire Rescue Training Center, it had a radio designation of Ambulance No. 99. The Ford was later assigned to Lucketts as Ambulance No. 10, when it was determined that an ambulance was needed in this part of the county. Ambulance No. 99 was painted white with a red stripe.

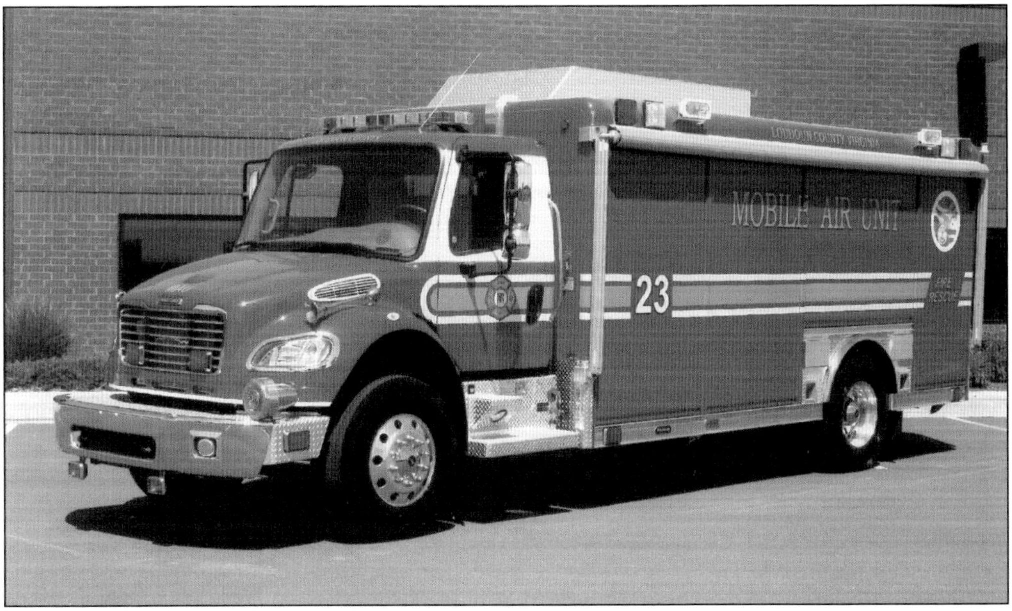

When the Pentagon was attacked on September 11, 2001, ample air supply for emergency workers was a problem. A grant provided seven mobile air units that were delivered to a variety of metropolitan Washington, D.C., fire departments. Loudoun County placed into service Mobile Air Unit No. 23 at Ashburn's Moorefield Station. The 2005 Freightliner with a Hackney body carries a large air compressor and approximately 100 extra air bottles, and it is equipped with a light tower and other equipment.

INDEX